About Island Press

Since 1984, the nonprofit organization Island Press has been stimulating, shaping, and communicating ideas that are essential for solving environmental problems worldwide. With more than 800 titles in print and some 40 new releases each year, we are the nation's leading publisher on environmental issues. We identify innovative thinkers and emerging trends in the environmental field. We work with world-renowned experts and authors to develop cross-disciplinary solutions to environmental challenges.

Island Press designs and executes educational campaigns in conjunction with our authors to communicate their critical messages in print, in person, and online using the latest technologies, innovative programs, and the media. Our goal is to reach targeted audiences—scientists, policymakers, environmental advocates, urban planners, the media, and concerned citizens—with information that can be used to create the framework for long-term ecological health and human well-being.

Island Press gratefully acknowledges major support of our work by The Agua Fund, The Andrew W. Mellon Foundation, Betsy & Jesse Fink Foundation, The Bobolink Foundation, The Curtis and Edith Munson Foundation, Forrest C. and Frances H. Lattner Foundation, G.O. Forward Fund of the Saint Paul Foundation, Gordon and Betty Moore Foundation, The Kresge Foundation, The Margaret A. Cargill Foundation, New Mexico Water Initiative, a project of Hanuman Foundation, The Overbrook Foundation, The S.D. Bechtel, Jr. Foundation, The Summit Charitable Foundation, Inc., V. Kann Rasmussen Foundation, The Wallace Alexander Gerbode Foundation, and other generous supporters.

The opinions expressed in this book are those of the author(s) and do not necessarily reflect the views of our supporters.

VITAL SIGNS

VITAL SIGNS

VOLUME 21

The Trends That Are Shaping Our Future

WORLDWATCH INSTITUTE

Michael Renner, *Project Director*

Ralph Albus   Alexander Ochs
Katie Auth   Grant Potter
Colleen Cordes   Janice Pratt
Robert Engelman   Tom Prugh
Maaz Gardezi   Michelle Ray
Milen Gonzalez   Michael Renner
Mark Konold   Laura Reynolds
Petra Löw   Cameron Scherer
Matt Lucky   Philipp Tagwerker
Evan Musolino   Sophie Wenzlau

Linda Starke, *Editor*
Lyle Rosbotham, *Designer*

ISLANDPRESS

Washington | Covelo | London

Contents

Technical Note

Units of measure throughout this book are metric unless common usage dictates otherwise. Historical data series in *Vital Signs* are updated in each edition, incorporating any revisions by originating organizations. Unless noted otherwise, references to regions or groupings of countries follow definitions of the Statistics Division of the U.N. Department of Economic and Social Affairs. Data expressed in U.S. dollars have for the most part been deflated (see endnotes for specific details for each trend).

Acknowledgments

As has been the case since the inception of this series in 1992, *Vital Signs* is produced by a group of committed and talented people, including Worldwatch staff researchers and outside authors. The articles in this book were first released on our companion online site, vitalsigns.worldwatch.org, between April 2013 and March 2014.

Many individual and institutional funders, as well as our generous Board, provide the support without which our work would not be possible. For their support during the past year of not just this volume but also our flagship publication, *State of the World*, as well as a range of other reports and projects, we are deeply grateful to a wide range of organizations and individuals.

We would like to extend our deepest appreciation to the following: Ray C. Anderson Foundation; The Asian Development Bank; Carbon War Room; Caribbean Community Secretariat; Climate and Development Knowledge Network; Del Mar Global Trust; Embassy of the Federal Republic of Germany in the United States; Energy and Environment Partnership with Central America; Estate of Aldean G. Rhymer; Garfield Foundation (discretionary grant fund of Brian and Bina Garfield); The Goldman Environmental Prize; The William and Flora Hewlett Foundation in partnership with Population Reference Bureau; Hitz Foundation; INCAE Business School; Inter-American Development Bank; International Climate Initiative of the German Federal Ministry for the Environment, Nature Conservation and Nuclear Safety; Steven C. Leuthold Family Foundation; The Low-Emissions Development Strategy–Global Partnership; MAP Royalty Inc. Sustainable Energy Fellowship Program; the National Renewable Energy Laboratory and the U.S. Department of Energy; Organization of American States; The Population Institute; Randles Family Living Trust; V. Kann Rasmussen Foundation; Renewable Energy Policy Network for the 21st Century; Serendipity Foundation; The Shenandoah Foundation; Town Creek Foundation; Turner Foundation; United Nations Foundation; United Nations Population Fund; Johanette Wallerstein Institute, Inc.; and Weeden Foundation.

For their financial contributions and in-kind donations, we would like to thank Ed Begley Jr., Edith Borie, Stanley and Anita Eisenberg, Robert Gillespie, Charles Keil, Adam Lewis, John McBride, Leigh Merinoff, MOM's Organic Market, Nutiva, George Powlick and Julie Foreman, Peter and Sara Ribbens, Peter Seidel, Laney Thornton, and three anonymous donors. Among the Worldwatch Board of Directors, we especially thank L. Russell Bennett, Mike Biddle, Edith Eddy, Robert Friese, Ed Groark, Nancy and Jerre Hitz, Isaac van Melle, David Orr, John Robbins, and Richard Swanson.

This edition was written by a team of 18 researchers. In addition to outside contributors Colleen Cordes and Petra Löw, a group of veteran Worldwatch researchers, former colleagues, and interns measured Earth's vital signs. They include Katie Auth, Robert Engelman, Milena Gonzalez, Mark Konold, Matt Lucky, Evan Musolino, Alexander Ochs, Grant Potter, Janice Pratt, Tom Prugh, Michael Renner, Michelle Ray, Laura Reynolds, Cameron Scherer, Philipp Tagwerker, and Sophie Wenzlau.

Vital Signs authors acknowledge help from experts who kindly offer data and insights on the trends we follow. We give particular thanks this year to Worldwatch Senior Fellow Erik Assadourian, Colin Couchman at IHS Automotive, and Ida Kubiszewski at the Institute for Sustainable Solutions at Portland State University.

Linda Starke, who has edited every edition in this series, ensures that all contributors follow not only grammar rules but also a consistent style. Once all the editing is completed, graphic designer Lyle Rosbotham works his magic to generate a visually pleasing layout and select a suitable cover image.

No less important are the people who work hard to oversee our work, manage the office, and ensure that our work is funded. We thank in particular Robert Engelman (who stepped down as Worldwatch President in early 2014 to once more focus on his first love—research, writing, and public speaking), Ed Groark (chairman of our Board and now interim President), Barbara Fallin (Director of Finance and Administration), Mary Redfern (Director of Institutional Relations), and Development Associates Courtney Dotson and Grant Potter.

During the past year, we bade farewell to a number of colleagues. Communications Manager Supriya Kumar left to pursue graduate studies. Grant Potter decided to pursue new professional opportunities, moving to California, as did Climate and Energy Research Associate Matt Lucky. Janice Pratt's term as an Atlas Corp Fellow from Liberia came to an end. Michelle Ray, who interned with Worldwatch's energy and climate team, as well as Laura Reynolds and Sophie Wenzlau, who worked on food and agriculture issues, also moved on to new challenges.

We would also like to thank our colleagues at Island Press. We benefit greatly from the ideas and other inputs provided by Maureen Gately, Jaime Jennings, Julie Marshall, David Miller, Sharis Simonian, and Brian Weese, and look forward to many years of a productive relationship with them.

Michael Renner
Project Director
Worldwatch Institute
1400 16th Street, N.W.
Washington, DC 20036
vitalsigns.worldwatch.org

Introduction:
A Global Disconnect

Michael Renner

Anyone still in doubt about the increasingly perilous state of our planet—and the implications for human civilization—need look no further than the most recent assessment by the Intergovernmental Panel on Climate Change, released in three reports between fall 2013 and spring 2014.[1]

Still, global studies can seem remote relative to particular local or national concerns. For the United States, the National Climate Assessment released in early May 2014 offers sobering specifics.[2] It underlines that climate change is already a reality and that warming in parts of the country could conceivably exceed 10 degrees Fahrenheit by the end of this century. The report seeks, as the *New York Times* put it, "to help Americans connect the vast planetary problem of global warming … to the changing conditions in their own backyards."[3]

Yet there is an ever more stark disconnect between what science tells us and what the political system is able and willing to do about it. All too often, governments are dragging their feet, and corporate CEOs see climate action as inimical to their profits. Indeed, as journalist and activist Naomi Klein has argued, "Climate change is a collective problem demanding collective action…. Yet it entered mainstream consciousness in the midst of an ideological war being waged on the very idea of the collective sphere."[4]

Energy policy across much of the globe can only be labeled as schizophrenic. It seems driven more by the ideology of endless growth than by concern for a livable future, more by corporate strategies than by the public interest, and more by considerations of supply security and geopolitics than by shared human needs.

As this edition of *Vital Signs* shows, global fossil fuel use is still growing. Coal, natural gas, and oil accounted for 87 percent of global primary energy demand in 2012. Coal—the dirtiest of the fossil fuels—may well become the dominant energy source by 2017.

Prevailing policies permit the pursuit of such forms of "extreme energy" as Arctic and deepwater deposits, tar sands, shale oil and gas unlocked through "fracking," and mountaintop-removal coal. Notwithstanding unresolved concerns about cost, safety, and waste storage, nuclear power is still advertised as a solution. Massive investments in extreme forms of energy lock society into an utterly unsustainable path. The investors are extremely powerful actors, and they have every interest in preventing society from choosing an alternative energy path.

The installed capacity of renewable energy worldwide is growing, that is true, and technologies like wind and solar photovoltaics are rapidly becoming more cost-competitive. But governmental support is still essential, and policy uncertainties—even reversals—have put a break on investments, while intensifying competition and trade disputes have caused disruptive realignments in the industry.

Not surprisingly, greenhouse gas emissions—and atmospheric concentrations—are hitting record levels. Emissions from fossil fuel combustion and cement production reached 9.7 gigatons in 2012 and were projected at 9.9 gigatons for 2013. Emissions from agriculture, too, are at peak levels. Energy efficiency gains to date are inadequate to stop these trends. Automobile and airplane fuel efficiencies are improving, for example, but not at a pace and scale able to offset the growing numbers of vehicles and planes or their expanding use.

Subsidies keep conventional forms of energy entrenched. Estimates of global payments in favor of fossil fuel production and consumption run anywhere from $523 billion to $1.9 trillion, depending on the definition applied. Even at the lower end of this range, spending dwarfs the amount made available in support of renewables.

Similarly, agricultural subsidies—some $486 billion in the top 21 food-producing countries in 2012—support factory farms that have colossal environmental footprints. They also often favor wealthy farmers and undermine farming in developing countries.

A failure to connect—to think and act across the boundaries of different disciplines and specializations—could well be diagnosed as human civilization's fundamental flaw in the face of growing and real threats. This is the case not just in the fields of energy and agriculture but also with regard to other key concerns, as this edition of *Vital Signs* suggests.

For instance, demographers issue rosy projections of future human life spans but fail to take into account that changing environmental conditions worldwide may well increase mortality in coming decades. Similarly, analysis of refugee and migration dynamics needs to take into account factors that have not traditionally been included. Population movements are increasingly due to complex, intertwined reasons, including environmental degradation, poverty and inequality, resource disputes, and poorly designed development projects. For a better understanding of the dynamics and for more productive discussions about possible policies, it is important that migration, refugee, and environmental experts work together closely.

Trends analyzed in this edition of *Vital Signs* also suggest another type of disconnect. At a time when climate change increasingly intersects with social and economic upheavals, disasters, and conflicts, governments continue to invest large sums in traditional forms of security policy. These troubling priorities mean that U.N. peacekeeping budgets of about $8 billion per year are not enough to cover even two days' worth of global military spending. Military spending also dwarfs aid flows. The $1,234 billion that high-income countries spent on military programs in 2012 is almost 10 times what they allocated for development assistance.

As climate change becomes ever more of a present-day reality rather than a far-away specter, priorities need to be revisited to boost the resilience of human societies and to minimize the stress factors that contribute to the world's conflicts and instabilities.

Energy Trends

A maintenance worker in the open nacelle of a wind turbine on Ascension Island in the South Atlantic

Lance Cheung

For additional energy trends, go to vitalsigns.worldwatch.org.

Fossil Fuels Dominate Primary Energy Consumption

Milena Gonzalez and Matt Lucky

Coal, natural gas, and oil accounted for 87 percent of global primary energy consumption in 2012 as the growth of worldwide energy use continued to slow due to the economic downturn.[1] The relative weight of these energy sources keeps shifting, although the change was ever so slight. Natural gas increased its share of global primary energy consumption from 23.8 to 23.9 percent during 2012, coal rose from 29.7 to 29.9 percent, and oil fell from 33.4 to 33.1 percent.[2] The International Energy Agency predicts that coal will replace oil as the dominant primary energy source worldwide by 2017.[3]

The shale revolution in the United States is reshaping global oil and gas markets. The United States produced oil at record levels in 2012 and was expected to overtake Russia as the world's largest producer of oil and natural gas combined in 2013.[4] Consequently, the country is importing decreasing amounts of these two fossil fuels, while using rising levels of its natural gas for power generation. This has led to price discrepancies between the American and European natural gas markets that in turn have prompted Europeans to increase their use of coal for power generation. Coal consumption, however, was dominated by China, which in 2012 for the first time accounted for more than half of the world's coal use.[5]

Natural gas consumption grew by 2.2 percent to 2,987 million tons of oil equivalent (mtoe) in 2012—more than triple the level in 1970.[6] (See Figure 1.) The largest increases in 2012 took place in the United States (an additional 27.6 mtoe), China (12.0 mtoe), and Japan (10.1 mtoe).[7]

Global natural gas production grew by 1.9 percent in 2012; the United States (with 20.4 percent of the total) and Russia (17.6 percent) are the dominant producers.[8] Other countries accounted for less than 5 percent each of global output.[9]

Estimated natural gas reserves grew by 0.7 percent in 2012 to 173,400 mtoe, principally due to increases in Iran (485 mtoe) and China (434 mtoe).[10] Reserves remain heavily concentrated in the Eurasia and Middle East regions; Iran, Russia, and Qatar alone account for 55 percent of the world's total.[11] The global reserves-to-production ratio now stands at 63.6 years, the length of time that current supplies would last if production were to continue at the same rate as in 2011.[12]

Natural gas prices continued to diverge globally in 2012.[13] (See Figure 2.) Prices have been on significantly different trajectories across the world since the economic meltdown in 2008 due to the U.S. shale gas revolution and increases in shale gas exports to Asian countries. While U.S. prices declined by 31 percent to $2.76 per million Btu (in current dollars), the prices increased by 5 percent in Europe and 14 percent in Japan.[14] Japan paid $16.75 per million Btu for liquefied natural gas (LNG) imports in 2012—a staggering six times more than the rate

Milena Gonzalez is the MAP Sustainable Energy Fellow at Worldwatch Institute. **Matt Lucky** was the sustainable energy lead researcher at the Institute.

paid by U.S. consumers—as it continues to rely on natural gas to make up for decommissioned nuclear power plants.[15] Japan is the dominant LNG importer, accounting for more than one-third of all LNG flows in 2012.[16]

While global LNG trade increased significantly during the past decade—growing by 120 percent since 2002—it declined for the first time by about 1 percent in 2012.[17] Australia, Indonesia, Malaysia, Nigeria, and Qatar dominate the market, accounting for about two-thirds of global LNG exports between them.[18]

For the second consecutive year, natural gas consumption fell in the European Union (EU), down by 2.3 percent in 2012.[19] This was the only region that experienced a decline. The causes include a struggling economy, increasingly large coal imports from Colombia and the United States, cheaper global coal prices, natural gas supply disruptions due to political unrest in parts of northern Africa, and natural gas producers' diversion of supplies to China and Japan.[20]

Consumption trends in the United States, which accounts for 22 percent of global natural gas use, have reflected recent price shifts among fossil fuels.[21] The country increased its use of natural gas by 4 percent during 2012.[22] Much of this growth occurred in the power sector, attributable to the declining price of natural gas relative to coal due to the U.S. shale gas revolution. In fact, electricity generation from natural gas equaled that from coal for the first time ever in April 2012.[23] But when the price of coal fell again, gas use in the power sector dropped by 14 percent during the first seven months of 2013 relative to the same period in 2012.[24]

In 2012, coal remained the fastest-growing fossil fuel globally, even though at 2.5 percent the increase in consumption was weak relative to the 4.4 percent average of the last decade.[25] (See Figure 3.) China increased its coal use by 6.1 percent.[26] India also saw significant increases in its coal consumption—9.9 percent in 2012.[27] Coal use by members of the Organisation for Economic Co-operation and

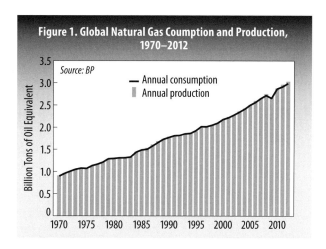

Figure 1. Global Natural Gas Coumption and Production, 1970–2012

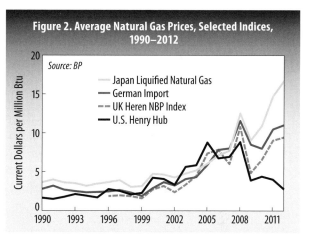

Figure 2. Average Natural Gas Prices, Selected Indices, 1990–2012

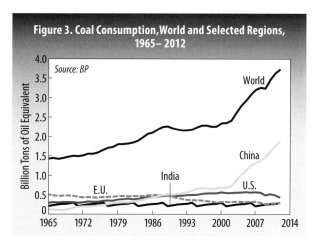

Figure 3. Coal Consumption, World and Selected Regions, 1965–2012

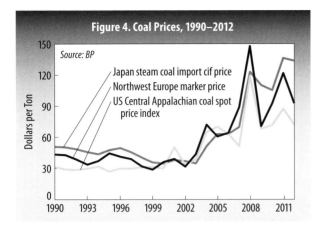

Figure 4. Coal Prices, 1990–2012

Source: BP

Japan steam coal import cif price
Northwest Europe marker price
US Central Appalachian coal spot price index

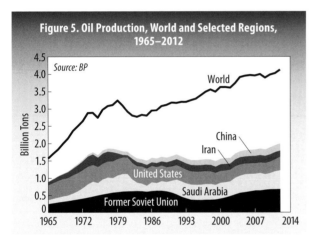

Figure 5. Oil Production, World and Selected Regions, 1965–2012

Source: BP

Development (OECD) declined by 4.2 percent, as an 11.9-percent decline in U.S. consumption outweighed increases of 3.4 percent in the EU and 5.4 percent in Japan.[28]

Global coal production grew by 2 percent in 2012.[29] A decline in U.S. output of 7.5 percent was more than offset by an expansion of 3.5 percent in China's production.[30] China now accounts for a dominant 47.5 percent of global coal production, followed by the United States (13.4 percent) and India (6 percent).[31] But the United States still holds the largest share of proven reserves, with 27.6 percent, followed by Russia with 18.2 percent, China with 13.3 percent, Australia with 8.9 percent, and India with 7 percent.[32]

Coal prices fell across all markets in 2012.[33] (See Figure 4.) After a relatively steady increase from early 2009 through mid-2011, European prices decreased from a peak $121.52 to $92.50 per ton.[34] Prices fell less drastically in other markets: from $87.38 to $72.06 in the United States and from $136.21 to $133.61 in the Pacific Basin.[35]

Oil remains the most widely consumed fuel worldwide, but at a growth rate of 0.9 percent, it is being outpaced by gas and coal for the third consecutive year.[36] The OECD's share declined to 50.2 percent of global consumption—the smallest share on record and the sixth decrease in seven years.[37] This reflects declines of 2.3 percent in U.S. consumption and of 4.6 percent in EU consumption.[38] By contrast, usage in China and Japan rose by 5.0 and 6.3 percent, respectively.[39]

Conversely, global oil production grew by more than twice as much as consumption—2.2 percent or 100.1 million tons in 2012.[40] (See Figure 5.) This was mainly due to a rise in U.S. output of 13.9 percent—the highest rate ever.[41] In comparison, Canada, China, and the former Soviet Union saw relatively small increases of 6.8, 2.0, and 0.4 percent, respectively.[42]

Conflicts in the Middle East and Africa disrupted oil supplies. Iran's oil production decreased by 16.2 percent or 33.3 million tons due to international sanctions.[43] Sudan's oil production declined by 81.9 percent, while Syria's was cut in half, dropping by 49.9 percent.[44] South Sudan, where most of the oil in the former Sudan is produced, gained independence from Sudan in mid-2011, but the new country still depends on Sudan for access to export pipelines and processing facilities.[45] A dispute over oil transit fees led South Sudan to shut down oil production in early 2012.[46] In April 2013, the country restarted production, yet several issues remain unresolved between the two nations.[47] And in Syria, civil war and sanctions have had a dramatic impact on the oil industry. Damage to its energy infrastructure and uncertainty over the outcome of the continued violence threaten Syria's energy sector and regional energy markets.[48]

These various disruptions were offset, however, by expanding output in many countries that belong to the Organization of the Petroleum Exporting Countries (OPEC). Libya's civil war had led to a collapse in production of 71 percent in 2011.[49] But in 2012, Libya saw an astonishing rebound of 215.1 percent, returning output to close to 2010 levels.[50] In addition, Saudi Arabia, the United Arab Emirates, and Qatar produced at record levels, with increases of 3.7, 1.6, and 6.3 percent, respectively.[51] Together, these three countries accounted for 19 percent of world output.[52] Iraq and Kuwait also saw significant increases in production, raising their combined share of global production to 7.4 percent.[53]

In 2012, crude oil prices peaked in March due to the reduction in Iranian oil exports, but they eased later in the year due to rising output in the United States, Libya, and other OPEC countries.[54] They also fell in the second quarter of 2012 due to concerns of global economic slowdown.

Global oil trade grew by 1.3 percent in 2012 to 2.7 billion tons, equivalent to 62 percent of worldwide output.[55] Declining exports from several regions were offset by larger shipments from Canada and North Africa. U.S. net oil imports fell by 37.4 million tons—to 22 percent below their peak in 2007.[56] China's net oil imports, on the other hand, grew by 30.4 million tons, accounting for 86 percent of the global increase.[57] According to projections by the U.S. Energy Information Administration, China will surpass the United States as the world's largest oil importer by 2014.[58]

Consumption of all fossil fuels will likely grow in the future. With increasing shale gas fracking and many countries' interest in displacing coal generation with natural gas due to the lower greenhouse gas emissions, natural gas use seems well poised to grow. Although some countries are trying to move away from coal use, the incredible coal consumption growth rates in China and India will likely make this the main energy resource in the next few years. Last, even if oil is eventually not the world's dominant energy resource, its use is expected to grow unless there is a fundamental change in the way the world fuels the transportation sector.

Nuclear Power Recovers Slightly, But Global Future Uncertain

Alexander Ochs and Michelle Ray

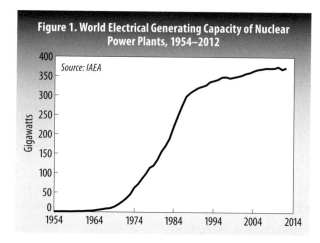

Figure 1. World Electrical Generating Capacity of Nuclear Power Plants, 1954–2012

Source: IAEA

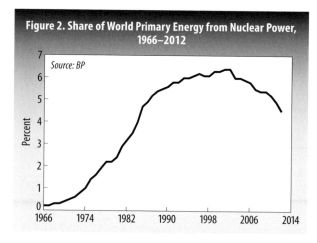

Figure 2. Share of World Primary Energy from Nuclear Power, 1966–2012

Source: BP

Alexander Ochs is director of the Climate and Energy Program at Worldwatch Institute. **Michelle Ray** was an intern at the Institute.

Global nuclear generation capacity increased by 4.2 gigawatts (GW), or 1.1 percent, to 373.1 GW in 2012.[1] (See Figure 1.) The number of operational reactors also increased in 2012 by two units, or 0.46 percent, to a total of 437 nuclear reactors worldwide.[2] The increases are net figures: three reactors with a total capacity of 1.3 GW were shut down in Canada and the United Kingdom, while three new plants in China and South Korea with a total capacity of just under 3 GW came online.[3] In addition, two Canadian reactors (with 772 megawatts (MW) each) returned to service after 15 years offline.[4]

Since 1987, expansion of the world's nuclear power generating capacity has slowed considerably. Just 75 GW were added over the last 25 years, compared with 296 GW during the preceding 25 years.[5] Indeed, nuclear power is the only mainstream energy technology that does not show rapid growth. Its share of the world's primary energy supply fell from 6.4 percent in 2001 and 2002 to just 4.5 percent in 2012, about the same share as in 1985.[6] (See Figure 2.)

Although nuclear power is dispersed widely across the globe, it is most heavily used in industrial countries. Of the 10 currently leading nuclear nations, 8 are established industrial countries, while China and South Korea are emerging industrial nations.[7] (See Figure 3.)

With 102.1 GW capacity and 104 reactors (including those under construction and in long-term shutdown), the United States remains the world's leading nuclear power.[8] Measured by its contribution to electricity generation, however, France is the leader.[9] Its 58 reactors contribute 75 percent of that country's electricity supply, compared with 19 percent in the United States.[10] (See Figure 4.)

Europe is the most reactor-saturated continent, with 170 plants—some 39 percent of the global total.[11] Following France's 58 reactors, Russia is second in Europe with 33 nuclear reactors in operation, 10 new ones under construction, and 26 in the planning phase.[12]

In Asia, China anticipates continued growth in its nuclear sector; it has 17 plants in operation, 29 under construction, and 38 in the planning phase.[13] South Korea's recent completion of 2 new reactors brings the total number of operational reactors there to 23, which generate 30.4 percent of total electricity production.[14] With 5 reactors currently under construction, the country aims to add a total of 11 new reactors and to increase its nuclear capacity to 43 GW by 2030.[15] In Japan, the share of electricity generation from nuclear power is unusually low for that country at the moment because many reactors were taken offline after the Fukushima disaster.[16]

Worldwide, construction began on seven new reactors during 2012, with total planned capacity of 6.9 GW, well short of the 15.8 GW of capacity that was started in 2010, when start-ups surged.[17] (See Figure 5.) China has led the world in capacity additions in recent years, and its 3.1 GW of new capacity accounted for 45 percent of global starts in 2012.[18] (See Figure 6.) Worldwide, some 67 nuclear reactors with a total capacity of 64.3 GW are being built.[19] However, 7 of these have now been under construction for more than 20 years, suggesting that their completion is doubtful.[20]

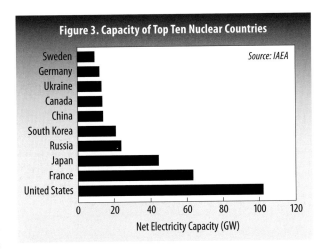

Figure 3. Capacity of Top Ten Nuclear Countries

Source: IAEA

Net Electricity Capacity (GW)

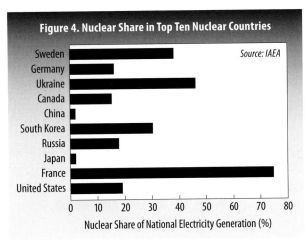

Figure 4. Nuclear Share in Top Ten Nuclear Countries

Source: IAEA

Nuclear Share of National Electricity Generation (%)

At the same time, decommissioning of aging nuclear power plants is at an all-time high, reflecting the aging of the global stock of nuclear power plants. The capacity of decommissioned plants reached 50 GW in 2011 and again in 2012, up from about 38 GW in both 2009 and 2010.[21] (See Figure 7.)

A number of European countries intend to reduce their nuclear power operations. Germany, with nine reactors, continues on its path to a full nuclear phaseout by 2022.[22] The Belgian government announced in July 2012 that all seven of the country's reactors will be shut down between 2015 and 2025.[23] French president François Hollande recently declared his intention to reduce reliance on nuclear power from 75 percent of electricity generation to 50 percent by 2025.[24] And the Swiss government released a national energy roadmap for public consideration that includes abandonment of its five nuclear power plants.[25]

Meanwhile, in the United States, the Nuclear Regulatory Commission suspended all reactor licensing decisions in 2012 "pending completion of a new waste

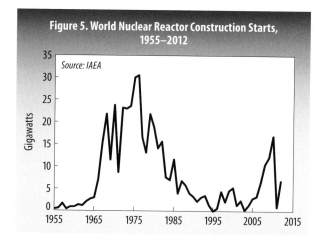

Figure 5. World Nuclear Reactor Construction Starts, 1955–2012

Source: IAEA

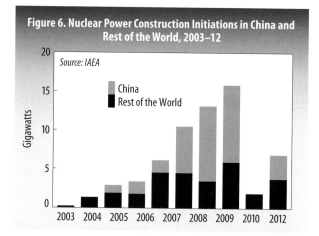

Figure 6. Nuclear Power Construction Initiations in China and Rest of the World, 2003–12

Source: IAEA

China
Rest of the World

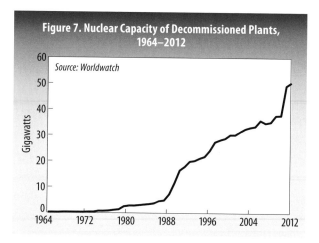

Figure 7. Nuclear Capacity of Decommissioned Plants, 1964–2012

Source: Worldwatch

confidence environmental impact statement and rule," with the Kewaunee plant (566 MW) and San Onofre-2 and 3 (1,070 MW and 1,080 MW, respectively) phased into permanent shutdowns in 2013.[26]

Several factors account for stagnancy in the nuclear sector. First is the high cost of nuclear power. Even at high levels of plant utilization, the levelized capital cost of nuclear energy (the cost of building and operating a plant over its operational life, divided by the energy generated) is 6–81 percent higher than the capital cost of other sources of energy for plants due to enter service in 2018.[27] (See Table 1.) In addition, the cost of nuclear power has increased dramatically over recent decades. In France, for example, the Fessenheim reactors were commissioned in 1978 at 1,070 euros per kilowatt electrical (kWe); Chooz 1 and 2 went online in 2000 at 2,060 euros per kWe; the delayed Flamanville 3 reactor is projected to produce energy at 3,700 euros per kWe when it is finalized.[28]

A second dampening influence is ongoing safety concerns. Despite stricter oversight provisions in some countries, public opposition to nuclear energy is still higher than approval in many places.[29] The accident at the Fukushima nuclear plant in Japan has tested public trust in nuclear institutions in that country. Ongoing leaks of radioactive water in 2012 and 2013 and radiation levels that are 18 times higher than initially announced by the operating company have shaken public confidence and led to the closing of all of Japan's nuclear plants in 2013.[30] Prior to the 2011 earthquake and tsunami, Japan generated 30 percent of its electricity from nuclear power—a share that was expected to increase to 40 percent by 2017.[31] Now, Japan's nuclear future is quite unclear.[32]

Meanwhile, less serious incidents have contributed to public nervousness about the technology. The list of these includes water leaks in Taiwan, water leaks and fires in France, cracks in reinforced concrete nuclear reactor structures and radiation leaks into the air in

Table 1. Capital Costs for Energy Plants, by Type, United States

Energy Type	Capital Cost	Cost Compared with Nuclear
	(2011 dollars per megawatt-hour)	(percent)
Natural Gas	15.8–44.2	19–53
Biomass	53.2	64
Coal	65.7	79
Wind	70.3	84
Geothermal	76.2	91
Hydropower	78.1	94
Nuclear	83.4	100

Source: EIA, "Levelized Cost of New Generation Resources in the Annual Energy Outlook 2013," January 2013.

the United States, as well as civilian break-ins, seawater leaks, and construction cracks in Sweden.[33] Growing security concern also relates to worries about possible terrorist attacks on nuclear installations and the proliferation of weapons-grade nuclear material. North Korea just proclaimed itself a nuclear state in 2012.[34]

Finally, the unresolved challenge of radioactive waste management dampens enthusiasm for nuclear power. Alternative solutions to geographical repositories are hampered by concerns about leakage of radioactive material. In August 2012, for example, evidence of a leak was found in one of the 177 underground double-shell waste tanks at the Hanford Nuclear Site in the United States—tanks that were thought to be stable and impervious.[35] To date, no solution to the nuclear waste challenge has proved safe and reliable for the extremely long time spans needed for nuclear radiation to abate.

Growth of Global Solar and Wind Energy Continues to Outpace Other Technologies

Matt Lucky, Michelle Ray, and Mark Konold

Global use of solar and wind energy continued to grow significantly in 2012. Solar power consumption increased by 58 percent, to 93 terrawatt-hours (TWh).[1] Use of wind power increased in 2012 by 18.1 percent, to 521.3 TWh.[2]

Although hydropower remains the world's leading renewable energy, solar and wind continue to dominate investment in new renewable capacity. They are quickly becoming the highest-profile renewable energies. And since most renewable energy policies worldwide focus on one or both of these resources, it is important to consider them together.

Global solar and wind energy capacities continued to grow even though new investments in these energy sources declined during 2012. Global investment in solar energy in 2012 was $140.4 billion, an 11 percent decline from 2011, and wind investment was down 10.1 percent, to $80.3 billion.[3] (See Figure 1.) But due to lower costs for both technologies, total installed capacities grew sharply.[4] (See Figure 2.)

Solar photovoltaic (PV) installed capacity grew by 41 percent in 2012, reaching 100 gigawatts (GW).[5] Over the past five years alone, installed PV capacity grew by a factor of 10 from 10 GW in 2007.[6] The countries with the most installed PV capacity today are Germany (32.4 GW), Italy (16.4 GW), the United States (7.2 GW), and China (7.0 GW).[7]

In 2012, installed capacity for concentrating solar power (CSP) reached 2.55 GW, with 970 megawatts (MW) alone added in 2012.[8] Spain (1.95 GW) and the United States (507 MW) dominate this market, while some growth has been seen in the North African nations of Algeria (25 MW), Egypt (20 MW), and Morocco (20 MW).[9]

Europe remains the dominant region in the solar sector, accounting for 76 percent of global solar power use in 2012.[10] Solar power use in Europe increased by 51.7 percent between 2011 and 2012.[11] Germany now accounts for 30 percent of the world's solar power consumption (a term that includes use of both solar PV and CSP).[12] (See Figure 3.)

Germany again led the world in added PV capacity in 2012, with 7.6 GW of new capacity.[13] Still, other countries in the region stood out, with PV capacity growing substantially in Denmark (up 2,872 percent), Bulgaria (544 percent), Greece (146 percent), and Austria (126 percent) in 2012.[14] Italy added the third most capacity of any country in the world in 2012 (3.4 GW), bringing its total installed capacity to 16.3 GW.[15] However, in June 2013, Italy reached the subsidy cap for its feed-in tariff (FIT) program, so future solar PV projects will no longer be eligible to receive FITs.[16]

Matt Lucky was the sustainable energy lead researcher at Worldwatch Institute. **Michelle Ray** was a Climate and Energy research intern. **Mark Konold** is the Caribbean program manager for the Climate and Energy Program.

Spain added 950 MW of installed CSP capacity in 2012, representing a growth of 95 percent.[17] Nevertheless, a retroactive change in FIT policies and a policy change allowing for taxation of all power producers will likely slow future growth in CSP there.[18]

The Asia-Pacific region now accounts for 17 percent of global solar use, leaving it behind only Europe.[19] Solar consumption grew by 69.5 percent in the region in 2012, and Japan (6.7 percent of the world total) and China (4.9 percent) are now among the top five global solar energy consumers.[20]

The region saw sustained growth in PV capacity, with significant increases in Malaysia (up 733 percent), India (435 percent), Thailand (140 percent), and China (100 percent) from 2011 to 2012.[21] Japan, largely as the result of a new FIT and the need for new energy sources after the disaster at Fukushima, installed the fifth-highest amount of PV capacity in the world in 2012, going from 5 to 7 GW.[22] Australia increased its installed PV capacity from 1.4 GW to 2.4 GW in 2012, largely due to the continuance of FITs by some state governments.[23]

Due to slowing global economic growth, decreasing demand, and oversupply, there were significant net losses in the Chinese PV industry, which supplies more than 50 percent of the world market.[24] Suntech—one of the world's largest PV manufacturers—recently defaulted on $541 million of bonds, generating fears that other PV manufacturers will default in the future.[25] Nevertheless, targets under China's twelfth Five-Year Plan include reaching 21 GW installed solar capacity by 2015 and 50 GW by 2020.[26]

The net losses for the Chinese PV industry are exacerbated by growing trade wars between China and both the European Union (EU) and the United States after they deemed that Chinese companies were dumping solar panels on their markets for prices below the production costs. The United States set anti-dumping duties ranging from 18.3 percent to 250 percent on Chinese manufacturers, and the EU levied an initial tariff of 11.8 percent in June, rising to 47.6 percent in August 2013 if there were no further negotiations.[27]

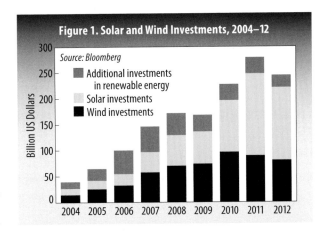

Figure 1. Solar and Wind Investments, 2004–12

Source: Bloomberg

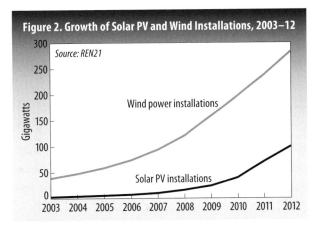

Figure 2. Growth of Solar PV and Wind Installations, 2003–12

Source: REN21

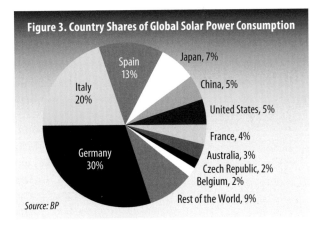

Figure 3. Country Shares of Global Solar Power Consumption

Source: BP

Solar power consumption increased by 123 percent in North America in 2012.[28] The United States alone accounted for 86 percent of the region's use of solar power.[29] North America added 3.6 GW of PV capacity in 2012, dominated by the United States (3.3 GW), helping the region reach a total installed capacity of 8.04 GW.[30]

The United States did not add any CSP capacity in 2012, but 1,300 MW of CSP are currently in construction and due to come online by 2014.[31] A 392-MW facility in the Mojave Desert is 75 percent complete and will become the world's largest operating CSP facility.[32]

In 2012, Africa's solar power consumption increased by 64.2 percent, while use in the Middle East increased by 62.3 percent.[33] The two regions still account for less than 1 percent of global solar consumption, but this number is poised to grow in the future.[34] Lower PV prices and the need for manufacturers to find new markets are driving new investments in PV in Africa and the Middle East.[35] In terms of CSP, the United Arab Emirates just began operation of a 100-MW plant in March 2013, while South Africa started construction on two plants (50 MW and 100 MW).[36]

South and Central America remain minor players in the global solar market despite having strong resource potential. In 2012, this region accounted for only 0.1 percent of global solar power consumption.[37]

Total installed wind capacity edged up in 2012 by 45 GW to a total of 284 GW, an 18.9 percent increase from 2011.[38] In keeping with recent years, the majority of new installed capacity was concentrated in China and the United States, which reached total installed capacities of 75.3 GW and 60 GW, respectively.[39]

The United States was the world's top wind market in 2012. Overall capacity increased 28 percent as the country added 13.1 GW, double the amount it added in 2011.[40] Increased domestic manufacturing of wind turbine parts, improved technological efficiency, and lower costs helped spur this increase, but the greatest catalyst was the threat of expiration of the federal Production Tax Credit (PTC)—which provides tax credits for kilowatt-hours produced by wind turbines—at the end of 2012.

Canada's installed wind capacity grew by 17.8 percent (935 MW), a drop from its 31.4 percent growth in 2011.[41] It was the ninth largest wind market in 2012, and Ontario and Quebec have plans to install another 1,500 MW combined in 2013.[42] Mexico passed the 1-GW mark in 2012 by adding 801 MW to its existing 600 MW.[43] It is now home to Latin America's largest wind power complex (360 MW) as the Oaxaca II, III, and IV projects were brought online in 2012.[44]

The EU remained the dominant region for wind power, as it passed an important milestone by installing 11.9 GW of new capacity to reach 106 GW, representing 37.5 percent of the world's market.[45] (See Figure 4.) Currently, wind accounts for 11.4 percent of the EU's total installed generation capacity.[46] Germany and Spain remained Europe's largest wind markets, increasing their total installed capacity to 31.3 GW and 22.8 GW, respectively.[47] The United Kingdom was third in new installations in 2012, at 1.9 GW, followed by Italy with 1.3 GW.[48]

Asia's 15.5 GW of new installed wind capacity, the highest of any region in 2012, ensured that it remains on the heels of the EU.[49] Total installed capacity increased to 97.6 GW in 2012.[50] And while China's 20.8 percent increase maintains the country's regional dominance, India showed respectable gains by adding 2.3 GW to bring its total installed capacity to 18.4 GW, a 14.5 percent increase over 2011.[51]

Political instability continued to slow growth in Africa and the Middle East, but installed capacity grew by 9.3 percent in 2012 compared with 2011's rate of 2.6 percent.[52] The region now has 1,135 MW installed.[53] Tunisia nearly doubled its capacity by adding 50 MW, and Ethiopia installed its first commercial-scale wind farm (52 MW).[54] Sub-Saharan Africa looks poised to lead the way in 2013 as South Africa continues making progress on over 500 MW of new wind power capacity.[55]

Latin America also saw significant growth in installed wind capacity, increasing by 53.7 percent from 2.3 GW to 3.5 GW.[56] Brazil increased its capacity by 75.3 percent, reaching a total installed capacity of 2.5 GW.[57] Argentina's 54 MW increase to 167 MW and Nicaragua's 40 MW increase to 102 MW total were also notable in the region.[58]

The wind turbine manufacturing industry was negatively affected by lower government support and overcapacity in 2012, leading to the cancellation of expansion plans and the scaling back of workforce and operations. A slowdown was originally expected in 2013, but the U.S. Congress extended the PTC until the end of the year, which bodes well for many American and European wind turbine and parts manufacturers that benefited from it.[59]

The top 10 wind turbine manufacturers accounted for 77 percent of market share in 2012.[60] (See Figure 5.) Asia and Europe had the most companies in the top 10 (5 and 4, respectively), although U.S.-based General Electric had the greatest share, largely due to developers trying to beat what they thought was the end of the PTC.[61]

Offshore wind capacity, mostly concentrated in northern Europe, increased by 1.3 GW (representing 2.9 percent of newly installed capacity in 2012) to bring the world's total offshore capacity to 5.4 GW.[62] The United Kingdom added 630 MW as part of the first phase of the London Array, reaching a total of 2.9 GW, followed by Denmark (0.9 GW), Belgium (0.4 GW), China (0.4 GW), Germany (0.3 GW), and Japan (0.25 GW).[63]

While policy uncertainties and changes will likely challenge the growth of

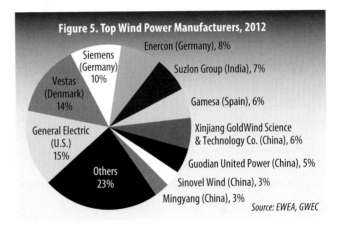

Figure 4. Total Installed Wind Capacity, 2010–12

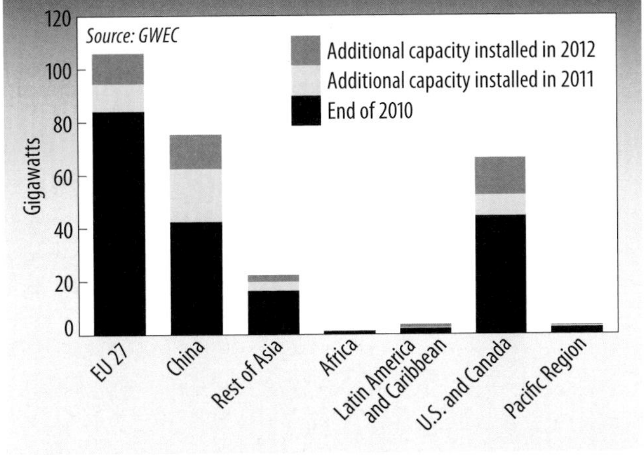

Source: GWEC

- Additional capacity installed in 2012
- Additional capacity installed in 2011
- End of 2010

Gigawatts

EU 27, China, Rest of Asia, Africa, Latin America and Caribbean, U.S. and Canada, Pacific Region

Figure 5. Top Wind Power Manufacturers, 2012

- Siemens (Germany) 10%
- Vestas (Denmark) 14%
- General Electric (U.S.) 15%
- Others 23%
- Enercon (Germany), 8%
- Suzlon Group (India), 7%
- Gamesa (Spain), 6%
- Xinjiang GoldWind Science & Technology Co. (China), 6%
- Guodian United Power (China), 5%
- Sinovel Wind (China), 3%
- Mingyang (China), 3%

Source: EWEA, GWEC

solar and wind in the future, these technologies are nonetheless well poised to grow. Declining solar technology prices, while challenging for current manufacturers, are helping solar to reach near grid-parity in many markets. With the decreasing cost of operating and maintaining onshore wind farms, onshore wind-generated power is already cost-competitive with conventional power energy sources in many markets.

Biofuel Production Declines

Tom Prugh

In 2012, the combined global production of ethanol and biodiesel fell for the first time since 2000, down 0.4 percent from the figure in 2011.[1] (See Figure 1.) Global ethanol production declined slightly for the second year in a row, to 83.1 billion liters, while biodiesel output rose fractionally, from 22.4 billion liters to 22.5 billion liters.[2] Biodiesel now accounts for over 20 percent of global biofuel production.[3]

Biofuels are a subset of bio-energy, which is energy derived from biomass (plant and animal matter) and which can range from manually gathered fuelwood and animal dung to industrially processed forms such as ethanol and biodiesel. Biomass can be used directly for heat, turned into biogas to produce electricity, or processed into liquid forms suitable as alternatives or supplements to fossil fuels for transport.[4] Bio-energy is regionally or locally important in many places around the world; traditional biomass is still used for cooking by 38 percent of people in the world, for example, while in parts of Africa and Asia, more than 90 percent of the populace relies on it.[5] In China and elsewhere in Asia, roughly 48 million biogas plants were built as of 2012 to support rural electrification.[6]

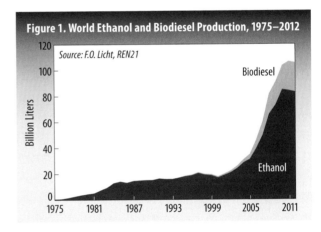

Figure 1. World Ethanol and Biodiesel Production, 1975–2012

Source: F.O. Licht, REN21

Biofuels for transport, essentially ethanol and biodiesel, account for about 0.8 percent of global energy use, 8 percent of global primary energy derived from biomass, 3.4 percent of global road transport fuels, and 2.5 percent of all transport fuels.[7] Ethanol is mainly derived from corn and sugarcane, while biodiesel comes from fats and vegetable oils.

The top five ethanol producers in 2012 were the United States, Brazil, China, Canada, and France.[8] (See Figure 2.) But the United States and Brazil accounted for 87 percent (61 percent and 26 percent, respectively) of the global total.[9] U.S. ethanol production totaled 50.4 billion liters, down about 4 percent from 2011; U.S. production depends mainly on corn as a feedstock, and corn prices rose in 2012 due to a severe summer drought in the Midwest.[10] As a result, in the autumn the United States briefly became a net importer of ethanol after nearly three uninterrupted years of net exports.[11] Brazil's production rose 3 percent to 21.6 billion liters, partly because of a drop in sugar prices.[12] The other top producers account for far smaller volumes; China's output, for instance, totaled 2.1 billion liters in

Tom Prugh is codirector of *State of the World 2013* and *State of the World 2014*.

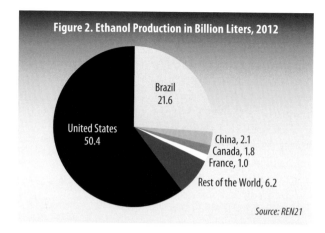

Figure 2. Ethanol Production in Billion Liters, 2012

United States 50.4

Brazil 21.6

China, 2.1
Canada, 1.8
France, 1.0

Rest of the World, 6.2

Source: REN21

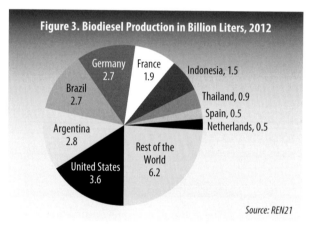

Figure 3. Biodiesel Production in Billion Liters, 2012

Germany 2.7
France 1.9
Indonesia, 1.5
Brazil 2.7
Thailand, 0.9
Spain, 0.5
Netherlands, 0.5
Argentina 2.8
Rest of the World 6.2
United States 3.6

Source: REN21

2012, while Canada's totaled 1.8 billion liters.[13] The European Union as a whole produced 4.6 billion liters of fuel ethanol in 2012.[14]

The United States also led in biodiesel production, with 3.6 billion liters, while Argentina took second place with 2.8 billion liters, and Germany and Brazil had roughly the same output at 2.7 billion liters apiece.[15] (See Figure 3.) China's biodiesel production, at only 200 million liters in 2012, lags far behind its output of fuel ethanol.[16] Several European nations produce biodiesel, and the European Union (EU) as a whole still accounted for 41 percent of global biodiesel output despite a decline of 7 percent in 2012.[17] Worldwide, biodiesel production grew at an average annual rate of 17 percent from 2007 through 2012, although the rate of growth slowed considerably.[18]

International trade in biofuels is significant, driven partly by supply and demand and partly by vagaries of trade policy. For example, the United States exported about 2.8 billion liters of ethanol in 2012, much less than in the previous year but still the second-highest export total ever.[19] Nearly one-third of that total, 893 million liters, went to Canada, with lesser amounts going to the United Kingdom, Brazil, and the Netherlands.[20] The same year, the United States imported 2.1 billion liters of ethanol, 83 percent of it from Brazil.[21] Likewise, of total U.S. biodiesel exports in 2012 of 486 million liters, 271 million liters (56 percent) went to Canada, even while U.S. imports of biodiesel from Canada totaled 67 million liters.[22] The United States also traded biodiesel imports and exports with Australia, the Netherlands, and Norway in 2012.[23]

Biofuel demand is strongly driven by blending mandates and supported by subsidies. Seventy-six states, provinces, or countries had such mandates on the books in 2012, up from 72 the previous year.[24] Global subsidies for liquid biofuels were estimated in 2012 to be well over $20 billion.[25] Mandates or targets have been established in 13 countries in the Americas, 12 in the Asia-Pacific region, and 8 in Africa.[26] In Europe, the EU-27 group of countries is subject to a Renewable Energy Directive (RED) that called for 5.75 percent biofuel content in transportation fuels in 2012.[27] The United States and China have established—and Brazil has already achieved—targets of between 15 and 20 percent no later than 2022; India has also mandated 20 percent ethanol by 2017.[28]

Whether these targets are stable and will be met is an open question, however. India, for example, is said to have an uneven record of meeting its own mandates.[29]

The EU's RED came under strong challenge in 2012 as a result of concerns over the effect that biofuel feedstock cultivation was having on food prices and changes in land use.[30] In response, the European Commission proposed limiting conventional biofuels (those derived from food crops) to a 5 percent share of all transport fuels.[31]

In the United States, the Environmental Protection Agency's mandate that petroleum refineries buy nearly 33 million liters of advanced cellulosic biofuels for blending in 2012 was thwarted by underproduction: less than 76,000 liters were produced that year.[32] A petroleum trade group later sued in federal court, successfully, to have the rule thrown out.[33] As with the challenge to the EU RED, the U.S. mandate—which requires biofuels made from non-food sources—reflects an attempt to reduce the non-energy impacts of biofuels policy. As U.S. corn-based ethanol production soared by a factor of seven between 2001 and 2010, corn prices and price volatility both increased, and additional lands were brought under cultivation around the world to boost food production.[34]

Global investment in biofuels equaled about $5 billion in 2012, down 40 percent from 2011; $3.8 billion of this was in industrial nations and $1.2 billion in developing ones.[35] Biofuels investment within the G-20 nations has declined every year from 2007 through 2012.[36] Despite this trend, some observers expect biofuel investment to rise; one forecast, for example, put 2023 revenues at $7.6 billion on investment of $69 billion over the decade, supported by continued blending mandates.[37] Where that investment goes will be influenced by the policy environment: in the United Kingdom, for instance, petroleum giant BP apparently aims to steer its biofuel investments to the United States and South America in response to the continued uncertainty within the EU over biofuel's impacts on land use and climate change.[38]

Policy Support for Renewable Energy Continues to Grow and Evolve

Evan Musolino

Throughout much of the world, government support remains essential for the growth of the renewable energy sector. Support policies for renewable energy technologies have increased dramatically over the last decade. Historically, policy design has evolved from an initial focus on supporting research and development in the 1970s and 1980s to today's focus on technology deployment and market development.[1] Starting in the mid-2000s, deployment-focused policies have been enacted at a rapid pace, growing from 48 countries with policies in place by mid-2005 to a total of 127 countries as of early 2013.[2] (See Figure 1.) This expansion occurred across geographic and economic boundaries. Developing and emerging economies accounted for less than one-third of all nations with policy support for renewable energy in 2005, but by 2013, they accounted for more than two-thirds.[3]

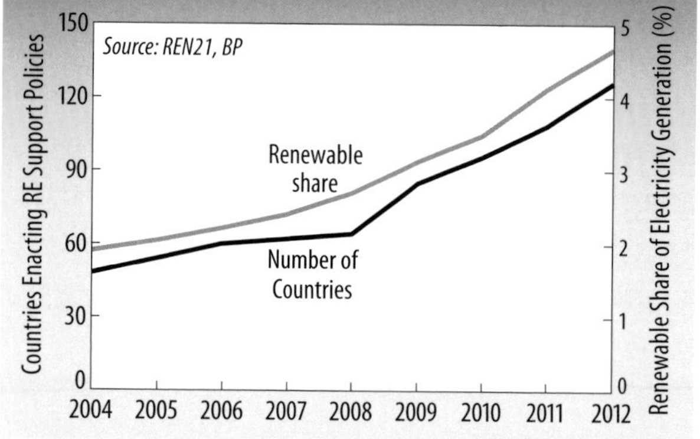

The majority of renewable energy support policies worldwide support electricity generation. Regulatory policies such as feed-in tariffs (FIT), net metering/billing, and renewable portfolio standards (RPS) or quotas have been developed to encourage the introduction of renewable energy technologies in the power sector.[4] The adoption of new policies has been matched by an increasing share of non-hydro renewable energy in the global electricity mix.[5]

Feed-in tariffs, under which a renewable generator receives a fixed payment over a set time period for electricity generated and fed into the grid, remain the most widely enacted form of policy support.[6] (See Figure 2.) The United States is credited with developing the precursor to the feed-in tariff in the Public Utility Regulatory Policies Act of 1978.[7] The first modern FIT was enacted in Germany in 1990, and by 2000 feed-in policies had been adopted by 14 countries.[8] Throughout the first decade of this century, the policy quickly gained momentum and became a primary tool for promoting renewable energy. An additional 54 countries enacted a FIT between 2000 and 2013.[9] Including state and provincial policies in Australia, Canada, India, and the United States, 99 FIT policies are now in place worldwide at the national or state/provincial level.[10] Municipal governments are also becoming more active in introducing FITs in cities.

Figure 1. Countries Enacting Renewable Energy Support Policies and the Share of Global Electricity Generation Supplied by Non-Hydro Renewables, 2004–12

Source: REN21, BP

Renewable share

Number of Countries

Evan Musolino is a research associate with the Climate and Energy Program at Worldwatch Institute and a contributor to REN21's *Renewables 2013 Global Status Report*.

Other policies promoting renewable energy have also seen a marked increase since the mid-2000s. (See Table 1.) Renewable portfolio standards or quotas for a specific required minimum share of renewable energy now exist in 76 countries, states, or provinces, up from 34 in 2004.[11]

A host of additional policies have been designed specifically to promote the uptake of renewable energies in the heating/cooling and transportation sectors, although their introduction stills lags behind policy activity in the electricity sector. At least 19 countries or states have enacted mandates for the use of renewable heat technologies.[12] Policies supporting renewable energy in the transportation sector through mandates and obligations are now in place in 51 countries at the national level.[13] Biofuel-blend mandates, a policy that requires a specific quantity of biofuels to be incorporated into transport fuel, are in place in 27 countries at the national level and in 27 states or provinces.[14]

Tax incentives are also being used by governments worldwide to spur developments in the renewable energy sector. As of early 2013, some 66 countries supported the renewable energy sector through their national tax codes.[15] Tax incentives can take many forms. Production tax credits, a main driver of the renewable energy sector in the United States, or investment tax credits allow for investments or stakes in renewable energy projects to be deducted from tax liabilities.[16] Many countries have also enacted measures to reduce or exempt specific taxes on renewable energy technologies—such as value added tax, sales tax, or import duties—in an effort to decrease costs of project development.[17]

As more policies have been enacted, regional diversity has greatly expanded. (See Figure 3.) Of countries enacting policies by mid-2005, most (58 percent) were found in Europe, followed by East Asia and the Pacific (21 percent), and by Latin America and the Caribbean (LAC) (13 percent).[18] At the time, regions such as sub-Saharan Africa and Central Asia had yet to develop any policy support mechanisms for renewable energies. By early 2013, the global picture had changed quite dramatically. Despite new policies in 15 European and Central Asian countries, that region's share of total policies enacted declined to slightly more than one-third of the global total.[19] In less than a decade, sub-Saharan Africa expanded

Figure 2. FITs Enacted at the National and State/Provincial Level, 2000–12

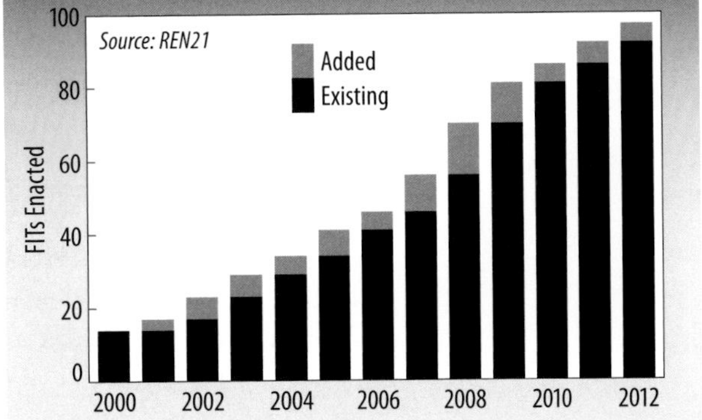

Source: REN21

Table 1. Countries Enacting Select National Renewable Energy Support Policies by Early 2013

Policy Type	Number of Countries
Feed-in Tariff	68
Tax Incentives	66
Biofuel Obligation/Mandate	51
Auctions/Tenders	44
Net Metering	31
Renewable Portfolio Standard	19
Renewable Heat Obligation/Mandate	14

Note: Number of countries indicates those where each policy has been identified at the national level.
Source: REN21, Renewables 2013 Global Status Report (Paris: 2013); IRENA, 2013.

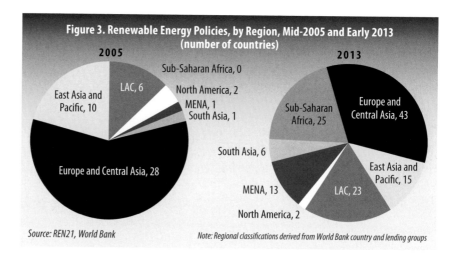

Figure 3. Renewable Energy Policies, by Region, Mid-2005 and Early 2013 (number of countries)

2005
- Sub-Saharan Africa, 0
- LAC, 6
- North America, 2
- MENA, 1
- South Asia, 1
- East Asia and Pacific, 10
- Europe and Central Asia, 28

2013
- Sub-Saharan Africa, 25
- Europe and Central Asia, 43
- South Asia, 6
- East Asia and Pacific, 15
- MENA, 13
- LAC, 23
- North America, 2

Source: REN21, World Bank

Note: Regional classifications derived from World Bank country and lending groups

from no renewable energy support policies to policies on the books in 25 countries, accounting for one-fifth of all nations enacting these policies worldwide.[20] A significant increase of 17 countries was also recorded in the LAC region, while 12 countries enacted policies for the first time in the Middle East–North Africa (MENA) region.[21]

The economic diversity of countries enacting support policies for renewable energy has also greatly expanded since the mid-2000s. (See Figure 4.) High-income economies accounted for 69 percent of all policy support by mid-2005, but by early 2013 this had declined to 30 percent.[22] The other economic groups each increased their shares by more than 10 percent.[23]

As the renewable energy sector continues to mature, policy makers face a host of new challenges. While the pace of countries adopting new renewable energy support policies has slowed somewhat in recent years, the sector has experienced a flurry of activity centered on revising existing policy mechanisms. Policy changes have been driven by a variety of factors, both positive and negative.[24]

Rapidly changing market conditions for technologies such as solar photovoltaics, where module costs have declined by 80 percent since 2008 and by 20 percent in 2012 alone, have dramatically reduced the level of support needed to make projects attractive to investors and feasible for project developers.[25] Simultaneously, the global economic slowdown left many countries with continuously tight national budgets, which has threatened support for the renewable energy sector. This combination of factors has led to a number of cuts to existing incentive programs. Across Europe, and to a lesser degree in other regions, cuts to FITs and other support policies have been undertaken in planned and unplanned ways.[26] The most damaging cuts have been made retroactively in such countries as Belgium, Bulgaria, the Czech Republic, Greece, and Spain, putting significant strain on the renewables industry.[27] Policy uncertainty is a key barrier constraining investment in the renewable energy sector as developers and investors seek stability.[28] Recent research suggests that risk reduction mechanisms such as longer-term contracts and grid connection guarantees

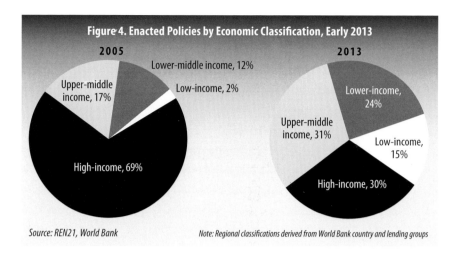

Figure 4. Enacted Policies by Economic Classification, Early 2013

2005

Lower-middle income, 12%

Upper-middle income, 17%

Low-income, 2%

High-income, 69%

2013

Lower-income, 24%

Upper-middle income, 31%

Low-income, 15%

High-income, 30%

Source: REN21, World Bank

Note: Regional classifications derived from World Bank country and lending groups

included within FIT schemes are more closely correlated than high FIT rates with greater deployment.[29]

Additional challenges to the sector have come from the introduction of taxes or tariffs on renewable energy installations or components.[30] Taxes on renewable energy were enacted in Bulgaria, Greece, and Spain in 2012.[31] Trade disputes centered on the international trade of renewable energy components have become prominent in recent years. Challenges have been brought to the World Trade Organization (WTO) against a number of countries for unfair trade practices, including subsidization, product dumping, and protectionist measures in policy design.[32] The WTO has recently ruled against the legality of the domestic-content requirement in Ontario's FIT, which includes a mandate that a specified share of an installed renewable energy system must come from Ontario in order to qualify for the incentive.[33] Based on this ruling, similar domestic-content requirements that exist elsewhere are now under scrutiny. Other countries have levied import duties on specific products to protect domestic interests.[34] The United States has placed duties on Chinese-manufactured solar modules and cells in response to separate subsidization and dumping claims, as well as tariffs on Chinese and Vietnamese wind towers.[35] China has also set tariffs targeting imported polysilicon, a key material used to manufacture solar panels, from suppliers in South Korea and the United States.[36]

New policy mechanisms are also being introduced. Auctions or tenders for renewable energy, also called public competitive bidding, are quickly emerging as an important tool, especially in developing and emerging economies. Under auction schemes, project developers bid the lowest price they would be willing to accept for a given quantity of renewable energy provided, either capacity or generation.[37] Certain countries, such as South Africa, are turning to auctions in place of more-traditional policy mechanisms such as FITs.[38] From 2009 to 2013, the number of countries using auctions to promote renewable energy technologies grew from 9 to 44.[39] High-income economies account for only 35 percent of the countries that have hosted renewable-energy auctions to date.[40]

As countries begin to have higher shares of renewable electricity in their national energy mix, policy makers need to address new challenges. Policy priorities are shifting from a need to incentivize market takeoff to ensuring favorable market conditions for the expanding renewables market and the seamless integration of renewable generation into grid networks.[41] Despite substantially different degrees of market maturation, renewable energy technologies continue to get support from government policy makers worldwide. This support is expected to continue, and to evolve, as the sector develops.

Phasing Out Fossil Fuel Subsidies

Philipp Tagwerker

Efforts to quantify global support for fossil fuels by the International Energy Agency (IEA), the Organisation for Economic Co-operation and Development (OECD), the International Monetary Fund (IMF), and a variety of nongovernmental organizations have generated a wide range of estimates. The amounts identified range from $523 billion to over $1.9 trillion, depending on the calculation and what measures are included.[1] What is clear is that the level of support has rebounded to 2008 levels following a dip in 2009–10 during the global financial crisis.[2]

Traditional calculations account for two kinds of energy subsidies. Production subsidies lower the cost of energy generation through preferential tax treatments and direct financial transfers (grants to producers and preferential loans). Consumption subsidies lower the price for energy users, usually through tax breaks or underpriced government energy services. While production subsidies predominate in OECD countries, consumption subsidies are favored in developing countries to reduce the burden on poor households, as poor people have to use a greater share of their income to buy fossil fuel products.

The IEA estimates coal, electricity, oil, and natural gas consumption subsidies in 38 developing economies at $523 billion in 2011.[3] Using a price-gap approach, the IEA figure includes subsidies that bring the price of fossil fuels below the international benchmark. Subsidies that lower the price just to the international level or slightly above it are not captured. In a parallel study by the OECD, support measures for the production and consumption of fossil fuels in its 24 member countries were inventoried.[4] Using a broader definition than the price-gap method (including direct budgetary transfers and tax expenditures), support for fossil fuels in OECD countries alone averaged $55–90 billion per year between 2005 and 2011.[5]

The lack of a clear definition of "subsidy" makes it hard to compare the different methods used to value support for fossil fuels, but the varying approaches nevertheless illustrate global trends. Fossil fuel subsidies declined in 2009, increased in 2010, and then in 2011 reached almost the same level as in 2008.[6] (See Figure 1.) The decrease in subsidies was almost entirely due to fluctuations in fuel prices rather than to policy changes. In developing countries, roughly $285 billion—more than 50 percent of all fossil fuel consumption subsidies—went to oil in 2011.[7] Natural gas consumption there received $104 billion in support.[8] Coal received only $3 billion in direct consumption subsidies in these countries, but another $131 billion went to public underpricing of electricity, much of which is generated from burning coal.[9] In industrial countries, using the broader definition of consumption subsidies, the support for oil was valued at roughly $38 billion in

Philipp Tagwerker is a research fellow at the Worldwatch Institute.

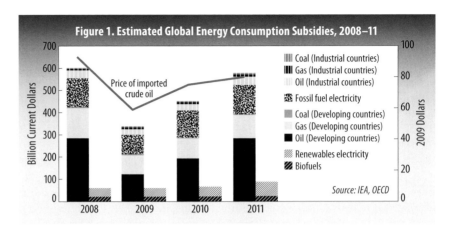

Figure 1. Estimated Global Energy Consumption Subsidies, 2008–11

2011.[10] Natural gas support in these countries totaled around $10 billion.[11] Coal was supported the least in industrial countries, receiving $7 billion in subsidies.[12]

Support for renewables is still small—$88 billion in 2011, compared with the support for fossil fuels estimated by the IEA and the OECD.[13] But it expanded by 33 percent in 2011, more than the 28 percent increase for fossil fuel subsidies.[14] Of the $88 billion support for renewables, two-thirds went toward electricity and the remaining one-third to biofuels.[15]

Producer subsidies other than the support measures captured in the OECD inventory were quoted at $100 billion in the June 2010 report to G-20 leaders from OECD, the IEA, the World Bank, and the Organization of the Petroleum Exporting Countries.[16] If further support measures such as export credit agencies (estimated at $50–100 billion annually) and the cost of securing fossil fuel supplies by protecting shipping lanes (estimated at $20–500 billion per year) are included, global subsidies, consumption, producer and other support measures amount to $1 trillion at the lower bound value.[17]

Externalities have so far not entered into the calculation of subsidies, although the additional costs associated with increased resource scarcity, the environment, and human health are enormous. Without factoring in such considerations, renewable subsidies cost between 1.7¢ and 15¢ per kilowatt-hour (kWh), higher than the estimated 0.1–0.7¢ per kWh for fossil fuels.[18] If externalities were included, however, estimates indicate fossil fuels would cost 23.8¢ more per kWh, while renewables would cost around 0.5¢ more per kWh.[19]

In a recent report, the IMF estimated support for fossil fuels on the basis of pre-tax and post-tax subsidies rather than classifying these as consumer or producer subsidies.[20] The IMF furthermore assumed a conservative $25 per ton in carbon dioxide (CO_2) damages, in light of widely varying estimates of between $12 and $85 per ton.[21] After removing tax breaks for energy products such as reduced value-added tax, the total amount of fossil fuel subsidies increased to at least $1.9 trillion, or 2.5 percent of global gross domestic product (GDP).[22]

These differing estimates and calculations highlight the need for increased transparency in governments' reports of fossil fuel subsidies. In its 2013 leadership communiqué, the G-20 reaffirmed its 2009 commitment to "phase out inefficient fossil fuel subsidies that encourage wasteful consumption over the medium term."[23] The governments reiterated the call for their own participation in a voluntary peer review process, as in the preceding years the number of countries opting out of reporting altogether had risen from two to six.[24]

Difficulties remain with the definition of the various elements of the G-20's commitment. Some member countries report far fewer subsidies than assessed by the OECD, the IEA, and the IMF, while others claim that their subsidies are not inefficient.[25] The discussion on subsidies is delicate, as issues of trade competitiveness, government sovereignty, and poverty alleviation come into play. In current discussions, it would help to decouple identifying and agreeing on what a subsidy constitutes from the commitment to phase them all out. Today the absence of harmonized subsidy data across countries is preventing the analysis of costs and distortions the subsidies place on the economy, a crucial first step in the process.[26]

An oft-cited argument for fossil fuel subsidies is the detrimental welfare impact of fuel price increases on low-income households. An IMF study of selected developing countries analyzed the distribution of subsidy benefits for five income levels of households.[27] It found that top-income households received, on average, six times more benefits than the supposed targeted group—the bottom one-fifth of households.[28] (See Figure 2.) By fuel type, the leakage of benefits to high-income groups is even more pronounced for gasoline (approximately 20 times more benefits) and liquefied petroleum gas (about 14 times more).[29]

For kerosene, low-income households' benefits are on par with the top one-fifth, but this still implies substantial leakage to top-income households, as people in this group do not rely on kerosene for cooking and lighting.[30] Fossil fuel subsidies are therefore an inadequate compensation measure for low-income households.

The reduction of fossil fuel subsidies often meets with political resistance, however, since poor households rely strongly on them. In early January 2012, Nigeria repealed one-third of its complete removal of gasoline subsidies after a nationwide strike, and Iran's second phase of subsidy reform has been put on hold for concerns of increasing inflation.[31]

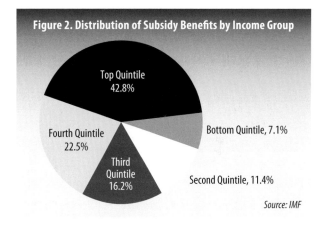

Figure 2. Distribution of Subsidy Benefits by Income Group

Top Quintile 42.8%

Bottom Quintile, 7.1%

Fourth Quintile 22.5%

Third Quintile 16.2%

Second Quintile, 11.4%

Source: IMF

Indonesia's experience is a good example of the political roller-coaster caused by a subsidy phaseout without adequate preparation. Following the rapid increase in fuel prices in the aftermath of the 1997 Asian financial crisis, widespread protests contributed to the end of President Suharto's rule.[32] Another attempt in 2005 to raise prices met far less resistance, as the reform of fossil fuel subsidies was coupled with unconditional cash transfers and other compensating measures for poor households.[33] Public information campaigns to raise awareness helped the Bantuan Langsun Tunai program reach 19.2 million households (35 percent of Indonesia's population).[34] In 2009, reforms and declining fuel prices managed to reduce the burden of fuel subsidies from 2.8 percent of GDP the previous year to 0.8 percent.[35]

Between 2010 and 2012, proposed increases in fuel prices were heavily slashed by the parliament, and the increase in international fuel prices caused Indonesia to spend 2.2 percent of GDP on its fuel subsidies in 2011.[36] Proposed fuel price increases in April 2012 and June 2013 both met with heavy protests, even though keeping them at past levels was going to jeopardize Indonesia's financial stability.[37] The lesson learned in Indonesia is that phasing out fossil fuel subsidies needs to be carefully planned, but communication campaigns and compensating welfare programs can alleviate resistance and actually provide the benefits that the subsidies were designed for in the first place.

Progress, albeit slow, has been made in some areas: Russia announced plans to increase regulated natural gas tariffs on the domestic market by 15 percent for all users from July 2013, and China announced that starting in March 2013, oil product prices would be adjusted every 10 working days to better reflect international prices.[38]

From an emissions perspective, 15 percent of global CO_2 emissions receive $110 per ton in support, while only 8 percent are subject to a carbon price, effectively nullifying carbon market contributions as a measure to reduce emissions.[39] Accelerating the phaseout of fossil fuel subsidies would reduce CO_2 emissions by 360 million tons in 2020, which is 12 percent of the emission savings that are needed in order to keep the increase in global temperature to 2 degrees Celsius.[40]

In summary, subsidy estimation and transparency initiatives need to be scaled up and consolidated. A common methodology and definition can aid in analyzing the distortions that subsidies cause in the economy and allow a better-informed dialogue to negotiate their phaseout. Furthermore, the same analysis could help alleviate the political resistance toward fuel price increases, as countries can be guided in their decision making and supported in the reallocation of funds.

Environment and Climate Trends

In the aftermath of Typhoon Bopha, Cateel, Davao Oriental, Philippines

Sonny Day

For additional environment and climate trends, go to vitalsigns.worldwatch.org.

Record High for Global Greenhouse Gas Emissions

Katie Auth

As climate negotiators, experts, and activists assembled in Warsaw, Poland, in November 2013, hoping to lay the groundwork for a global climate agreement in 2015, newly released data revealed continued growth in emissions of atmospheric carbon dioxide (CO_2) and other major greenhouse gases (GHGs), as well as a shifting geographic distribution of emissions.

According to the Global Carbon Project, CO_2 emissions from fossil fuel combustion and cement production reached 9.7 gigatons of carbon (GtC) in 2012, with a ±5 percent uncertainty range.[1] This is the highest annual total to date—and it is 58 percent higher than emissions in 1990, the year often used as a benchmark for emissions trends.[2] Coal (43 percent) and oil (33 percent) accounted for the majority of these emissions, with natural gas (18 percent), cement production (5 percent), and flaring (1 percent) making up the remainder.[3] (See Figure 1.) The Global Carbon Project's projection for 2013 was 9.9 ± 0.5 GtC, indicating an increase of approximately 2 percent.[4]

Recent U.S. government and World Bank moves to limit international financing for new coal projects signal a desire to shift away from this particularly carbon-intensive resource.[5] For now, however, coal remains a major driver of CO_2 emissions. Although it made up 43 percent of global emissions in 2012, coal accounted for 54 percent of the increase that year, reflecting in part rising coal use in countries currently undergoing energy sector transitions.[6] Coal-related emissions increased, for example, in Germany (4.2 percent) and Japan (5.6 percent)—both of which are phasing out nuclear power plants.[7] Oil, gas, and cement accounted for 18 percent, 21 percent, and 6 percent, respectively, of the global increase in 2012.[8] (See Figure 2.)

Although CO_2 is the primary greenhouse gas emitted through human activities, it is not the only one with significant warming effects. Other major long-lived greenhouse gases include methane (CH_4), nitrous oxide (N_2O), and chlorofluorocarbons (CFCs), particularly CFC-12 and CFC-11. Each gas's contribution to climate change depends on such factors as the length of time it remains in the atmosphere, how strongly it absorbs energy, and its atmospheric concentration.[9]

Primarily as a result of fossil fuel combustion as well as deforestation and land use change, the mean atmospheric concentration of CO_2 stood at approximately 393.9 parts per million (ppm) in 2012, an increase of more than 40 percent since 1750 and of 24 percent since the Scripps Institution of Oceanography began keeping records in 1959; the institution's initial 2013 average from January through September was 396.2 ppm.[10] (See Figure 3.) Scientists have suggested that the CO_2 concentration will need to be reduced to at least 350 ppm if we hope to maintain

Katie Auth is a research associate in the Climate and Energy Program at Worldwatch Institute.

a climate similar to that which has supported human civilization to date.[11] Atmospheric CO_2 concentration increased by 2.2 ppm in 2012 alone, which exceeds the average annual increase over the past 10 years.[12] The slightly lower annual increase in 2011 (1.84 ppm) has been attributed in part to unusually high levels of land carbon uptake in 2010 and 2011 associated with La Niña weather patterns.[13]

The atmospheric concentrations of other major long-lived greenhouse gases have also increased. Globally averaged CH_4 concentration increased by 6 parts per billion (ppb) in 2012, reaching a new high of approximately 1,819 ppb.[14] CH_4 is the third most abundant greenhouse gas in the atmosphere, after CO_2 and water vapor; on a per molecule basis, however, methane has a more potent warming impact on the climate.[15] Although atmospheric CH_4 levels declined during 1983–99 and remained relatively constant during 1999–2006, they have been increasing since 2007.[16] Attributing this renewed growth rate to specific factors is difficult, but some analysts suggest that the causes include warm Arctic temperatures and increased precipitation in the tropics.[17] The measured concentration in 2012 represents an overall increase of approximately 169 percent since 1750.[18]

The increasing concentration of these gases in the atmosphere has significant warming impacts. The combined heating effect (or "radiative forcing") of major long-lived GHGs (particularly CO_2, CH_4, N_2O, CFC-12, and CFC-11), as measured by the Annual Greenhouse Gas Index of the U.S. National Oceanic and Atmospheric Administration, reached 1.32 in 2012.[19] This indicates a 32 percent increase since 1990 and a 63 percent increase since 1980.[20] Because of its abundance, CO_2 has played a particularly large role in this, accounting for approximately 80 percent of radiative forcing by long-lived GHGs between 1990 and

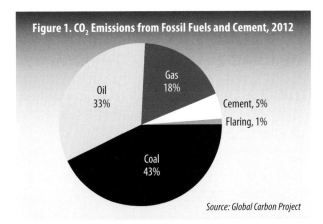

Figure 1. CO_2 Emissions from Fossil Fuels and Cement, 2012

Source: Global Carbon Project

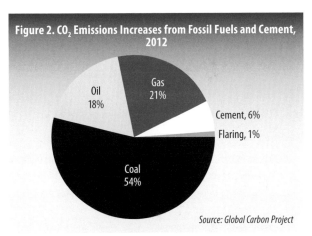

Figure 2. CO_2 Emissions Increases from Fossil Fuels and Cement, 2012

Source: Global Carbon Project

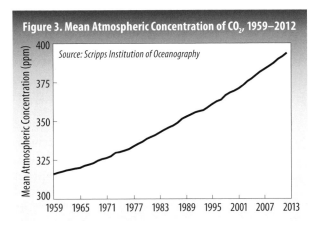

Figure 3. Mean Atmospheric Concentration of CO_2, 1959–2012

2012.[21] (See Figure 4.) A significant decline in the global use of CFCs has limited net radiative forcing. If the 1987 Montreal Protocol had not regulated these ozone-depleting gases, it is estimated that they would have contributed additional

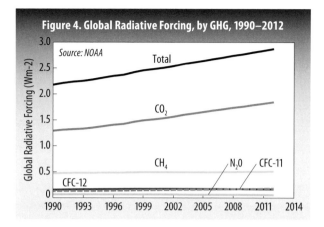

Figure 4. Global Radiative Forcing, by GHG, 1990–2012

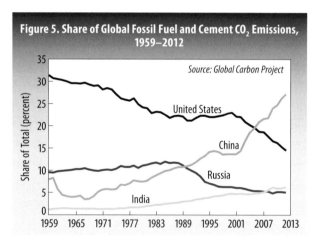

Figure 5. Share of Global Fossil Fuel and Cement CO_2 Emissions, 1959–2012

radiative forcing equivalent to more than half that of CO_2 between 1990 and 2012.[22]

Although in 2010 the parties to the United Nations Framework Convention on Climate Change agreed that the increase in average global temperature since the pre-industrial period must be kept below 2 degrees Celsius, many projections now put the climate on track for warming that is significantly above that. The Global Carbon Project foresees a 3.2–5.4 degrees Celsius "likely" increase in temperature.[23] A World Bank report projects an approximate 20 percent likelihood of exceeding a 4 degrees Celsius increase by 2100 if current mitigation commitments and pledges are not fully implemented.[24]

Emissions data from 2012 also highlight the shifting geographical and historical complexity that makes international negotiations so contentious. Global distribution of emissions in 2012 looks very different than it did in 1990, when the Kyoto Protocol was established. At that time, industrial countries accounted for 62 percent of emissions; by 2012, that figure had dropped to 37 percent, reflecting rapid industrialization and development in emerging economies and shifting patterns in production and consumption.[25]

Annual growth rates highlight the rapidly growing climate importance of large, emerging economies. China's emissions rose by 5.9 percent in 2012, an increase that accounted for 71 percent of that year's global increase.[26] The United States and Australia, although both still major emitters, experienced reductions of 0.05 percent and 11.6 percent, respectively.[27] In 2012, the top four emitters of CO_2 were China (2,625.7 million tons of carbon, or MtC), the United States (1,396.8 MtC), India (611.2 MtC), and the Russian Federation (491.8 MtC).[28]

Fossil fuel and cement emissions in these countries have been on dramatically different trajectories, making for contentious relations between U.S. and Chinese climate negotiators.[29] (See Figure 5.) Although China now accounts for more emissions than the United States does, the U.S. cumulative total since 1959 of 68,908 MtC is still 71 percent higher than that of China, the next largest emitter at 40,279.22 MtC.[30] In 2012, per capita emissions in China equaled those of the 28 members of the European Union at 1.9 tons—still far below the U.S. figure of 4.4 tons but significant in highlighting changing distribution patterns.[31]

Although international climate negotiations have traditionally focused on the role and responsibility of nation states, new analyses point to the significant role of corporate entities in emitting greenhouse gases. Overall, investor-owned corporations have been responsible for 21.71 percent of CO_2 and CH_4 fossil fuel and cement emissions since 1750, with state-owned corporations responsible for an additional 19.84 percent, highlighting potential new ways to frame responsibility for climate mitigation.[32]

As climate negotiators, experts, and activists left Warsaw and began to work on forging a global deal in Paris in 2015, they had to grapple with these changing complexities.

Agriculture and Livestock Remain Major Sources of Greenhouse Gas Emissions

Laura Reynolds

Global greenhouse gas (GHG) emissions from the agricultural sector totaled 4.69 billion tons of carbon dioxide (CO_2) equivalent in 2010, an increase of 13 percent over 1990 emissions.[1] (See Figure 1.) By comparison, global CO_2 emissions from transport totaled 6.76 billion tons that year, and emissions from electricity and heat production reached 12.48 billion tons.[2]

Growth in agricultural production between 1990 and 2010 outpaced growth in emissions by a factor of 1.6, demonstrating increased energy efficiency in the agriculture sector.[3]

The U.N. Food and Agriculture Organization (FAO) maintains country-specific data for annual GHG emissions from agriculture. These data measure nitrous oxide, carbon dioxide, and methane—the three most common gases emitted in agriculture. Methane is generally produced when organic materials—such as crops, livestock feed, or manure—decompose anaerobically (without oxygen).[4] Methane accounts for 49.8 percent of total agricultural emissions.[5] Enteric fermentation—the digestion of organic materials by livestock—is the largest source of methane emissions and of agricultural emissions overall.[6]

Nitrous oxide is a by-product generated by the microbial breakdown of nitrogen in soils and manures.[7] Nitrous oxide production is particularly high in cases where the nitrogen available in soils exceeds that required by plants to grow, which often occurs when nitrogen-rich synthetic fertilizers are applied.[8] Nitrous oxide is responsible for 36.3 percent of agricultural GHG emissions.[9]

Finally, carbon dioxide is released from soils when organic matter decomposes aerobically (with oxygen). The largest source of CO_2 emissions within agriculture is the drainage and cultivation of "organic soils"—soils in wetlands, peatlands, bogs, or fens with high organic material.[10] When these areas are drained for cultivation, organic matter within the soil decomposes at a rapid rate, releasing carbon dioxide.[11] This process accounts for around 14 percent of total agricultural GHG emissions.[12]

Figure 2 shows the top six sources of global GHG emissions from the agricultural sector. Emissions from each activity varied by region, reflecting different agricultural production trends around the world. For example, while enteric fermentation accounted for 29 percent of emissions in both North America and Asia in 2010—the lowest percentage of all regions—it was the source of 61 percent of South America's agricultural emissions, reflecting that continent's world leadership in cattle production.[13] Similarly, rice cultivation was responsible for 17 percent of Asia's total emissions in 2010 but no more than 3 percent of emissions in every other region—indicating Asia's dominance of global rice output.[14]

Laura Reynolds was a staff researcher in Worldwatch's Food and Agriculture Program.

Emissions from enteric fermentation rose by 7.6 percent worldwide between 1990 and 2010, but regional variation was again high.[15] (See Figure 3.) At 51.4 percent and 28.1 percent, respectively, Africa and Asia saw their emissions increase, while those in Europe and Oceania fell by 48.1 percent and 16.1 percent, respectively.[16] Europe's significant reduction in emissions parallels the decline in its beef production between 1990 and 2010, but it may also reflect increased use of grains and oils in cattle feed instead of grasses.[17]

Adding oils or oilseeds to feed can assist digestion and reduce methane emissions.[18] However, a shift from a grass-based to a grain- and oilseeds-based diet often accompanies a shift from pastures to concentrated feedlots, which in turn, unfortunately, has a range of negative environmental consequences, such as water pollution and high fossil fuel consumption.[19] Aside from reducing overall livestock populations, there is therefore no clear pathway to climate-friendly meat production from livestock.

Manure deposited and left on pastures also contributes to global nitrous oxide emissions because of its high nitrogen content. When more nitrogen is added to soil than is needed, soil bacteria convert the extra nitrogen into nitrous oxide and emit it into the atmosphere—a process called nitrification.[20] Emissions from manure on pastures were highest in Asia, Africa, and South America, accounting for a combined 81 percent of global emissions from this source.[21] (See Figure 4.) In these three regions, emissions from this source increased by an average of 41.5 percent between 1990 and 2010, reflecting an increase in range-based livestock populations; in all other regions, these emissions either decreased or stagnated.[22]

Carbon dioxide emissions from cultivated organic soils were the source of approximately 14 percent of total agricultural emissions.[23] When organic soils are drained for agriculture, the exposure of accumulated organic matter to oxygen causes rapid decomposition and

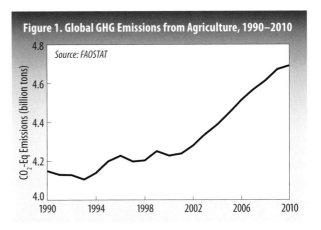

Figure 1. Global GHG Emissions from Agriculture, 1990–2010

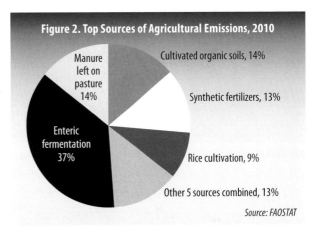

Figure 2. Top Sources of Agricultural Emissions, 2010

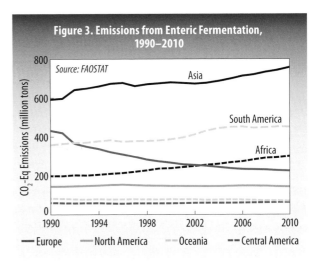

Figure 3. Emissions from Enteric Fermentation, 1990–2010

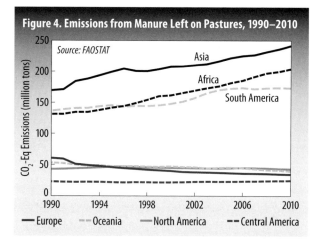

Figure 4. Emissions from Manure Left on Pastures, 1990–2010

Source: FAOSTAT

CO$_2$-Eq Emissions (million tons)

— Europe – – Oceania — North America – – Central America

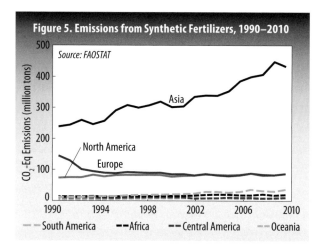

Figure 5. Emissions from Synthetic Fertilizers, 1990–2010

Source: FAOSTAT

CO$_2$-Eq Emissions (million tons)

– – South America – – Africa – – Central America – – Oceania

destroys a stable carbon sink, releasing carbon dioxide into the atmosphere.[24]

Asia was responsible for 53.8 percent of global emissions from this source, indicating the deforestation and clearing for agricultural land that has occurred in many tropical Southern and Southeast Asian countries.[25] In fact, four out of the top five countries with the highest emissions from cultivated organic soils were in Asia—Indonesia contributed 278.7 million tons of carbon dioxide from this source, Papua New Guinea 40.8 million tons, Malaysia 34.5 million tons, and Bangladesh 30.6 million tons.[26]

Deforestation also contributes significantly to global carbon dioxide emissions. While the main purpose of forest clearing is to create agricultural land, there are few reliable data on forestland cleared specifically for agriculture on the global level. Between 1990 and 2010, some 143.7 million hectares of forestland were cleared, releasing 65.9 million tons of carbon dioxide.[27]

Emissions from synthetic fertilizers rose 37.6 percent from 1990 to 2010, but this increase took place almost entirely in Asia.[28] (See Figure 5.) This trend could indicate Asia's widespread abandonment of traditional, manure-based field fertilization practices in favor of synthetic fertilizers, to keep up with skyrocketing production of cereal crops. In 2010, after rising 81.3 percent since 1990, Asia's emissions from synthetic fertilizers were nearly twice as high as these emissions from all other regions combined.[29] China alone was responsible for over one-third of the world's synthetic fertilizer emissions in 2010.[30] These high emissions correlate with Asia's rapid increase in crop production in recent decades: it jumped 72.4 percent since 1990, including an 81.1 percent spike in China's production.[31]

Since many crops, such as corn, need high levels of nitrogen to thrive, farmers opt to use nitrogen-based synthetic fertilizers.[32] But applying these fertilizers often causes the same soil-nitrification as manure deposits on pastures do, producing nitrous oxide.[33] Nitrogen can also leach into surface and groundwater, endangering wildlife and public health.[34]

Applying fertilizer more efficiently, precisely, and at times when plants can absorb it can significantly reduce nitrous oxide production.[35] Efficient fertilizer use can also reduce the fossil fuel emissions that result from the production of synthetic fertilizers (which are not included in FAO's estimate of agricultural emissions). Planting fallow fields with nitrogen-fixing legume crops—such as soybeans, alfalfa, and clover—can also naturally rebuild nitrogen in soils.[36]

Rice cultivation is the fourth most significant source of agricultural emissions worldwide. Most of the world's rice is grown in flooded fields, or paddies, causing organic matter such as rice husks to decompose without oxygen, which produces methane.[37] Emission trends from rice cultivation mirror the area of rice harvested in Asia between 1990 and 2010: emissions rose 7.3 percent and the area harvested rose 8.2 percent, with considerable variation within this period.[38] (See Figure 6.) Asia accounted for 88.7 percent of total emissions and 88.9 percent of the global area of rice harvested.[39]

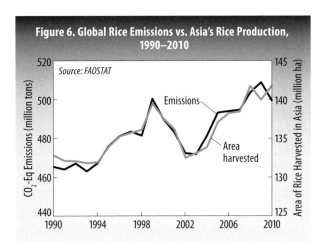

Figure 6. Global Rice Emissions vs. Asia's Rice Production, 1990–2010

Source: FAOSTAT

The Intergovernmental Panel on Climate Change (IPCC) suggests that draining rice paddies once or several times during the growing season can reduce methane emissions—but it notes that this might not be possible in many parts of the world due to insufficient rainfall or irrigation.[40] Keeping soils as dry as possible in the non-rice-growing season and incorporating organic material into fields during this season can lower methane production.[41]

All these data on GHGs make the huge proportion of emissions that is attributable to livestock production clear. Emissions from manure left on pastures by livestock, manure applied to soils, and manure treated in management systems account for 23 percent of total agricultural emissions.[42] In fact, FAO points out that when emissions from enteric fermentation and from cropland devoted to feed production are added to this category, to encompass all emissions directly related to livestock production, the proportion is over 80 percent.[43] This leaves emissions related to the direct human consumption of food crops at just under 20 percent.[44] Consumers can help lower agricultural emissions by reducing their consumption of meat and dairy products. This would help stabilize or lower livestock populations, lessen the pressure to clear additional land for livestock, and reduce the proportion of grain that is grown for livestock feed instead of for direct human consumption.

Aside from reducing livestock's contribution to climate change, farmers and landowners have numerous opportunities for mitigation, many of which offer environmental and even economic co-benefits. Growing trees and woody perennials on land can sequester carbon while simultaneously helping to restore soils, reduce water contamination, and provide beneficial wildlife habitat.[45] Reducing soil tillage can also rebuild soils while lowering GHG emissions.[46] Some practices can even result in increased income for farmers—"cap-and-trade" programs allow farmers to monetize certain sequestration practices and sell them, while government programs like the U.S. Conservation Reserve Program pay farmers to set aside some of their land for long-term restoration.[47] As the IPCC notes, many mitigation practices use existing and accessible technologies and can be implemented immediately.[48]

Natural Catastrophes in 2012 Dominated by U.S. Weather Extremes

Petra Löw

In 2012, there were 905 natural catastrophes worldwide—and 93 percent of these events were weather-related disasters.[1] This figure was about 100 above the 10-year annual average of 800 natural catastrophes.[2] In terms of overall and insured losses ($170 billion and $70 billion, respectively), 2012 did not follow the records set in 2011 and could be defined as a moderate year on a global scale.[3] But the United States was seriously affected by weather extremes: it accounted for 69 percent of overall losses and 92 percent of insured losses due to natural catastrophes worldwide.[4] (See Figure 1.)

Of the 905 documented loss events, 45 percent were meteorological events (storms), 36 percent were hydrological events (floods), and 12 percent were climatological events such as heat waves, cold waves, droughts, and wildfires.[5] The remaining 7 percent were geophysical events—earthquakes and volcanic eruptions.[6] This distribution deviates somewhat from long-term trends, as between 1980 and 2011 geophysical events accounted for 14 percent of all natural catastrophes.[7]

Some 37 percent of natural catastrophes occurred in Asia, 26 percent in the United States, 15 percent in Europe, 11 percent in Africa, 6 percent in Australia/Oceania, and 5 percent in South America.[8] This breakdown was approximately in line with the long-term average from 1980 to 2011.[9] Yet the trends of weather-related catastrophes show considerable regional differences. The largest increases

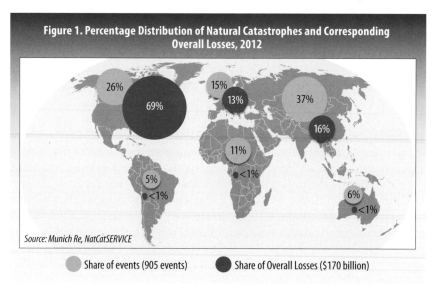

Figure 1. Percentage Distribution of Natural Catastrophes and Corresponding Overall Losses, 2012

Source: Munich Re, NatCatSERVICE

Share of events (905 events)　　Share of Overall Losses ($170 billion)

Petra Löw is a geographer and a consultant at Munich Reinsurance Company with a focus on natural catastrophe losses.

over the last 30 years occurred in North America (including Central America and the Caribbean), Asia, and Australia, while the smallest increases happened in Europe and South America.[10]

Deaths during natural catastrophes in 2012 stood at 9,600—substantially below the 10-year annual average of 106,000.[11] The percentage distribution of fatalities in the four recognized event categories—meteorological, hydrological, climatological, and geophysical events—also differed in 2012. Only 7 percent of the deaths were caused by geophysical events, compared with the long-term average since 1980 of 40 percent.[12] In contrast, 93 percent of the fatalities in 2012 were caused by weather-related events.[13]

Almost 30 percent of all fatalities were due to just five events: In January, a cold wave affected the eastern part of Europe and caused 530 fatalities.[14] From July until October, floods in Nigeria killed 431 people.[15] An earthquake in Iran in August killed 306 people.[16] In September, severe floods affected Pakistan, with 455 fatalities.[17] And by the end of the year, Typhoon Bopha struck the Philippines and caused 1,100 deaths, making that storm the year's most devastating natural catastrophe in terms of lives lost.[18]

Some two-thirds of the global overall losses and 92 percent of the insured losses in 2012 were due to weather-related events in the United States.[19] Hurricane Sandy, the summer-long drought in the Midwest, and severe storms with tornadoes accounted for $100 billion of the overall losses.[20] The insurance industry covered $58 billion of this.[21] These losses were the second highest overall and insured losses since 1980 in the United States.[22] (See Figure 2; overall and insured losses are adjusted for inflation.) The most expensive year in the United States was 2005, when Hurricane Katrina hit the coast of Mississippi and the city of New Orleans.[23] During the rest of 2012, overall losses occurred in Europe (13 percent) and the Asia-Pacific region (17 percent).[24] The insured losses for these two regions were well below the average and accounted for 8 percent of the total.[25]

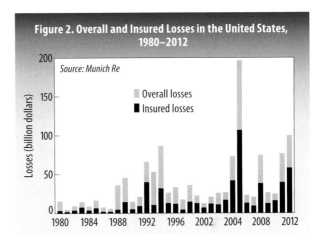

Figure 2. Overall and Insured Losses in the United States, 1980–2012

Source: Munich Re

- Overall losses
- Insured losses

A breakdown of the losses in the four event categories reveals a number of substantial deviations from the long-term average. Around 59 percent of overall losses in 2012 are attributable to storms, compared with the long-term average of 39 percent.[26] Earthquakes accounted for 12 percent of overall losses, but that was only half the 1980–2011 average.[27] With regard to insured losses, a particularly striking feature in the climatological events category was that droughts accounted for 28 percent.[28] This is well above the long-term average of 7 percent and was due to the severe drought that primarily afflicted the U.S. Midwest during the year, causing immense agricultural losses.[29] All in all, this drought was responsible for overall losses of $20 billion.[30] In addition, severe drought conditions affected Russia, Ukraine, Kazakhstan, and some parts of Southern Europe.[31]

Table 1. Costliest Natural Catastrophes in 2012					
Date	Event	Region	Overall Losses	Insured Losses	Fatalities
			(million dollars)	(million dollars)	
24–31 Oct	Hurricane Sandy, storm surge	Caribbean, United States	65,000	30,000	210
June–September	Drought, heat wave	United States	20,000	15,000–17,000	100
20/29 May	Earthquake	Italy	16,000	1,600	18
21–24 July	Floods	China	8,000	180	151
2–4 March	Severe storm, tornadoes	United States	5,000	2,500	41
28–29 April	Severe storm	United States	4,600	2,500	1
28 June–2 July	Severe storm	United States	4,000	2,000	18
25–30 May	Severe storm, hailstorm	United States	3,400	1,700	
10–24 May	Floods, landslides	China	2,500		127
3–27 September	Floods	Pakistan	2,500		455

Source: Munich Re, NatCatSERVICE database.

Table 1 lists the 10 costliest natural catastrophes in 2012 in terms of overall losses as registered in the Munich Re NatCatSERVICE database. Six of the top 10 occurred in the United States.[32]

The Caribbean was affected by three hurricanes (Isaac, Rafael, and Sandy) and some minor floods and flash-flood events.[33] Earthquakes occurred in Mexico, Guatemala, and Costa Rica.[34] Volcanic activity with property damage was registered in Nicaragua, Guatemala, and Mexico, causing only minor damage.[35]

The Atlantic tropical storm season began with Alberto on 19 May 2012, followed by Beryl on 26 May, which preceded the official start of the hurricane season on 1 June.[36] A total of 19 tropical storms occurred in 2012, with 10 storms reaching hurricane strength.[37] The number of tropical storms and hurricanes is well above the long-term average. Seven storms made landfall, with four of them battering the United States (Beryl, Debby, Sandy, and Isaac).[38] Hurricane Sandy, the costliest event in the year in terms of both overall and insured losses, caused losses in eight different countries, and 15 states were affected in the United States.[39] The overall losses from Sandy reached $65 billion, of which $30 billion was covered by insurance.[40]

As noted, just 5 percent of natural catastrophes in 2012 affected South America.[41] And 86 percent of these were weather-related events.[42] This was in line with the distribution of the previous year.[43] Only a few events, like floods in Colombia, Paraguay, and Peru and winter damages in Argentina, caused relevant property damages, totaling $600 million in overall losses.[44]

In 2012, some 15 percent of global natural catastrophes occurred in Europe.[45] A series of earthquakes in Italy's Emilia Romagna province proved exceptionally expensive, with overall losses of $16 billion.[46] With insured losses of $1.6 billion, this series became the insurance industry's costliest earthquake loss ever in Italy.[47] Many of the region's buildings, including historic monuments, were destroyed.[48] In January, two winter storms hit Western and Northern Europe and caused overall losses of $1 billion.[49] The southern part of Europe was affected by drought conditions, especially parts of Italy, Spain, Portugal, Croatia, Bosnia and Herzegovina, Serbia, and Slovenia, with a total of $3.8 billion in overall losses.[50] Also, severe dry conditions affected parts of Russia and Kazakhstan; the agricultural losses reached $600 million for Russia alone.[51]

Some 11 percent of natural catastrophes happened in Africa, slightly above the long-term average of 9 percent.[52] All of these events in 2012 were weather-related. A severe flood hit Nigeria from July to October and caused overall losses of $500 million.[53] It was one of the five deadliest events in 2012, as 360 people died.[54] The southern part of Africa was hit several times by tropical storms. The main losses were caused by Cyclones Giovanna and Funso and by Tropical Storm Dando and totaled $300 million in overall losses.[55]

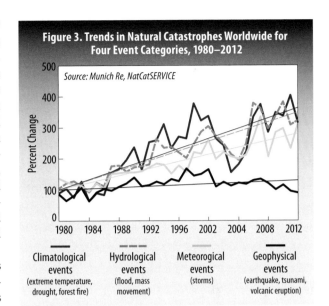

Figure 3. Trends in Natural Catastrophes Worldwide for Four Event Categories, 1980–2012

Source: Munich Re, NatCatSERVICE

Climatological events (extreme temperature, drought, forest fire)

Hydrological events (flood, mass movement)

Meteorogical events (storms)

Geophysical events (earthquake, tsunami, volcanic eruption)

The Asia-Pacific region accounted for 43 percent of all natural catastrophes worldwide in 2012.[56] After 2011, when major earthquakes in Japan and New Zealand featured prominently in global loss statistics, 2012 was a moderate year in the region in terms of losses. As noted earlier, only 17 percent of overall losses came from the Asia-Pacific region—well below the long-term average of 45 percent since 1980.[57] Most of the events were floods and storms. Asia was hit by major floods again, especially in China. Torrential rainfall in mid-June caused heavy losses in northeastern and southeastern China. Overall losses amounted to $8 billion.[58] Floods affected Australia as well, although the losses were relatively low compared with previous years and reached just $500 million.[59] In New Zealand, only a few natural catastrophes happened, mostly severe storms and flash floods.[60]

In terms of trends in natural catastrophes over the long term, Munich Re's Nat-CatSERVICE documents about 21,000 loss events worldwide from 1980 to 2012.[61] Some 13 percent of these events were geophysical, such as earthquakes, tsunamis, and volcanic eruptions.[62] But 87 percent were weather-related: meteorological (39 percent), hydrological (35 percent), or climatological (13 percent) events.[63] Since 1980, geophysical events have been more-or-less stable, whereas weather-related events have increased 2.8- to 3.6-fold. (See Figure 3.)[64]

Transportation Trends

Electric car plugged in at a public charging station in Amsterdam

For additional transportation trends, go to vitalsigns.worldwatch.org.

Automobile Production Sets New Record, But Alternative Vehicles Grow Slowly

Michael Renner and Maaz Gardezi

World auto production set yet another record in 2012 and may have risen even higher during 2013. According to London-based IHS Automotive, passenger-car production rose from 62.6 million in 2011 to 66.7 million in 2012, and it may have reached 68.3 million in 2013.[1] (See Figure 1.) When cars are combined with light trucks, total light vehicle production rose from 76.9 million in 2011 to 81.5 million in 2012 and was projected to total 83.3 million in 2013.[2]

Just four countries—China, the United States, Japan, and Germany—produced 53 percent of all light vehicles worldwide, and the top 10 countries accounted for 76 percent.[3] (See Figure 2.) At 18.2 million vehicles, China produced almost as many as the next two countries—the United States and Japan—combined.[4] Germany's and South Korea's production is declining slightly, while that of India, Brazil, Mexico, Canada, and Thailand is gaining.[5]

Automobiles are major contributors to air pollution and greenhouse gas emissions. Fuel efficiency standards are compelling manufacturers to produce cleaner cars that emit less carbon per kilometer driven. As of 2012, Japan, the European Union (EU), and India have the lowest limits, at between 128 and 138 grams of carbon dioxide per kilometer (g CO_2/km), whereas allowable emissions are in the range of 198–205 grams in Australia, Mexico, and the United States (for light-duty vehicles, LDVs).[6] (See Figure 3.)

The standards now on the drawing boards are intended to bring average per-vehicle emissions down to 95 grams in the EU and 105 grams in Japan (both by 2020), to 93 grams for U.S. cars and 109 grams for U.S. light-duty vehicles (by 2025), and to 153 grams in South Korea (by 2015).[7] India and China are studying limits for 2020 of 113 and 117 grams, respectively, and Mexico is considering 173 grams for 2016.[8]

At present, individual automobile manufacturers in Europe are producing vehicles that emit from 126 to 161 g CO_2/km.[9] Light vehicles purchased in the United States averaged emissions of 232 g CO_2/km in 2012.[10] Thus manufacturers need to make substantial progress to meet emission limits over the next decade or so.

Alternative vehicle propulsion technologies are slowly becoming more prominent, driven by a desire to reduce dependence on petroleum and the need to reduce air pollutants and greenhouse gas emissions. Alternatives include so-called hybrid vehicles that use both a conventional internal combustion engine and an electric motor, as well as a variety of electric vehicles (EVs), such as plug-in hybrid electric vehicles (PHEVs), battery electric vehicles (BEVs), and fuel cell electric vehicles.

The cumulative number of hybrid vehicles sold worldwide as of early 2013 was about 6.3 million.[11] Between late 1997 and March 2013, Toyota—the leading

Michael Renner is a senior researcher at Worldwatch Institute and codirector of *State of the World 2014*. **Maaz Gardezi** was a research fellow at the Institute.

hybrid manufacturer—sold nearly 5.13 million hybrids worldwide, or 81 percent of the global total.[12] In 2012, hybrids accounted for 14 percent of the company's global sales and 40 percent of its sales in Japan.[13] Honda, which introduced its first hybrid model in 1999, surpassed the 1 million cumulative sales mark in September 2012.[14]

Japan is the world's largest market for hybrids, with Toyota alone selling 678,000 vehicles there in 2012, more than double the 316,000 in 2011.[15] The United States is second, with 435,000 hybrids sold in 2012.[16] This compares with about 100,000 hybrids sold in Europe that year.[17] In the United States, hybrids and EVs combined are slowly increasing their share of total sales. They accounted for 3.8 percent of total U.S. passenger-car sales in the first five months of 2013, up from 2.4 percent in 2010.[18] But EV sales numbers are still dwarfed by those for hybrid cars.[19] (See Figure 4.)

Although more than doubling from 45,000 in 2011 to 113,000 in 2012, worldwide electric-vehicle production is still a minuscule 0.2 percent compared with that of conventional automobiles.[20] Sales are almost evenly split between PHEVs and BEVs (about 55,000 and 57,000 vehicles, respectively), while those of fuel cell EVs are very small.[21] In the United States, Canada, and the Netherlands, plug-ins dominate sales, while in Japan, China, France, Norway, Germany, and the United Kingdom, battery EVs are more popular.[22] (See Figure 5.)

By the end of 2012, the global EV fleet was estimated at more than 180,000—just 0.02 percent of all passenger cars in the world.[23] Table 1 shows EV fleets and charging stations in member countries of the Electric Vehicle Initiative (EVI), which brings together most of the leading EV markets.[24] The largest EV fleets are currently found in the United States, Japan, France, and China. Relative to population, however, the Netherlands, Japan, France, and Denmark have the largest number of EVs.[25] Norway, which is not an EVI member, currently has the fifth-largest EV fleet in the world, at

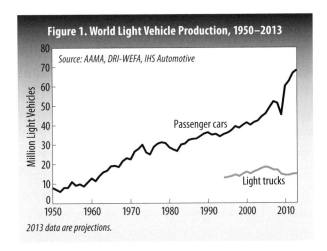

Figure 1. World Light Vehicle Production, 1950–2013

Source: AAMA, DRI-WEFA, IHS Automotive

2013 data are projections.

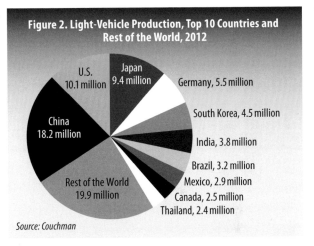

Figure 2. Light-Vehicle Production, Top 10 Countries and Rest of the World, 2012

Source: Couchman

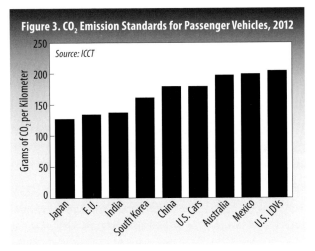

Figure 3. CO$_2$ Emission Standards for Passenger Vehicles, 2012

Source: ICCT

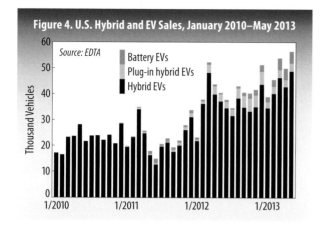

Figure 4. U.S. Hybrid and EV Sales, January 2010–May 2013

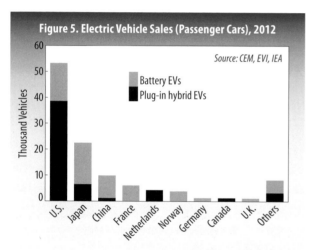

Figure 5. Electric Vehicle Sales (Passenger Cars), 2012

10,000 vehicles, and by far the greatest density: 200 EVs per 100,000 inhabitants.[26]

The development of a network of charging stations is proceeding unevenly in individual countries. The highest density of charging points, relative to the existing EV fleet, is found in Denmark, followed by Sweden, Spain, Italy, China, Portugal, and India.[27]

The major drivers of EV development include government targets for future EV sales and fleets, public investments in research and development for improved batteries and other technologies, and stricter fuel efficiency requirements. The targets issued by governments around the world add up to annual sales of 7.2 million vehicles by 2020 and a fleet of 24 million vehicles on the road the same year.[28] The bulk of this figure is accounted for by the 15 member states of the Electric Vehicle Initiative—the United States, China, Japan, India, South Africa, and 10 European countries—which aim to have annual sales of 5.9 million EVs by 2020 and a fleet of 20 million vehicles.[29] EVI governments provided more than $8.7 billion in R&D funds for this effort between 2008 and 2012.[30] In addition, more than $3 billion was spent on consumer incentives and close to $1 billion on EV infrastructure.[31]

Batteries are a key component of alternative propulsion vehicles. Lithium-ion batteries are already used heavily in consumer electronics and are growing in importance for the motor-vehicle industry. Challenges that need to be addressed include improving energy densities, reducing charging times, extending their cycle life, and reducing costs.[32]

Globally, battery production—for portable devices, grid storage, and vehicles—accounts for 29 percent of total lithium use.[33] Presently only 9 percent of lithium used in manufacturing batteries goes toward developing electric and hybrid car batteries.[34] But it is believed that in the years leading up to 2025, batteries will be the most important demand driver for lithium, increasing by 10 percent each year.[35] Hybrid and electric car batteries will provide the strongest impetus for this growth; their future production is projected to rise at a compound annual growth rate of 27.3 percent, at least until 2025.[36] The overall market for lithium-ion batteries in light-duty transportation is expected to grow from $1.6 billion in 2012 to $22 billion in 2020.[37] This is similar to a Boston Consulting Group projection of $25 billion by then.[38]

China, Japan, South Korea, and the United States are currently the major

Table 1. Electric Vehicle Fleet and Infrastructure, EVI Countries, 2012

	EV Fleet	EVs per 100,000 People	Charging Stations/ Points	Vehicle-to-Station Ratio
United States	71,174	22.7	15,192	4.7
Japan	44,727	35.1	5,009	8.9
France	20,000	31.4	2,100	9.5
China	11,573	0.9	8,107	1.4
United Kingdom	8,183	12.9	2,866	2.9
Netherlands	6,750	40.4	3,674	1.8
Germany	5,555	6.8	2,821	2.0
Portugal	1,862	17.6	1,350	1.4
Italy	1,643	2.7	1,350	1.2
India	1,428	0.1	999	1.4
Denmark	1,388	24.8	3,978	0.3
Sweden	1,285	13.5	1,215	1.1
Spain	787	1.7	705	1.1
Finland	271	5.0	2*	135.5
Total, Selected Countries	176,626	5.2	49,368	3.6

* This does not take into consideration electric block heaters, which are also used for charging.
Note: No data available for South Africa.
Source: Adapted and calculated from CEM, EVI, and IEA, Global EV Outlook. Understanding the Electric Vehicle Landscape to 2020 (April 2013), p. 4.

lithium-ion battery manufacturers for hybrid and electric vehicle applications, and a well-established electronics manufacturing industry gives East Asian countries a competitive advantage.[39] According to Roland Berger Strategy Consultants, the overall market share in 2015 for lithium-ion cell manufacturing for plug-in electric vehicles will likely be dominated by Japanese firms such as Panasonic, AESC, Toshiba, and GS Yuasa (with a combined share of 57 percent), followed by South Korea's Samsung and LG Chem (18 percent) and by China (9 percent, primarily due to its acquisition of the U.S. company A123 Systems Inc.).[40]

The U.S. lithium-ion battery manufacturing industry is still in an early stage of development. Current federal policies to promote electric vehicles include $2.4 billion in grants to battery makers, $3.1 billion in loans to automakers, and $1.8 billion in the form of tax credits to consumers for purchasing electric cars.[41] The Department of Energy (DOE) is backing research into new battery technologies and low-cost manufacturing methods. At the moment, however, the slowdown in demand for electric and hybrid vehicles has negatively affected lithium battery makers, causing insolvency for two of the major DOE grantees, A123 Systems Inc. and EnerDel.[42]

Air Transport Keeps Expanding

Michael Renner

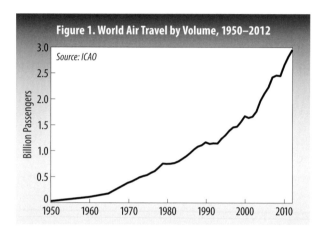

Figure 1. World Air Travel by Volume, 1950–2012

Source: ICAO

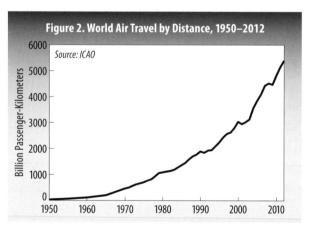

Figure 2. World Air Travel by Distance, 1950–2012

Source: ICAO

In 2012, the number of people traveling on airplanes reached 2,957 million, which was 4.7 percent more than the previous year.[1] (See Figure 1.) Although this figure includes a substantial number of people who travel multiple times during the year, it is equivalent to 42 percent of the world's population.[2] The number of passengers is up 95-fold from 31 million in 1950, when flying was a luxury few could afford, and it is triple the 960 million passengers in 1986, when air travel was already quite common.[3]

The average length of a flight doubled from 903 kilometers in 1950 to 1,816 kilometers in 2000, but it has not changed much since then and stood at 1,827 kilometers in 2012.[4] Longer flights and expanding passenger numbers generated a strong expansion of total passenger kilometers (pkm) traveled—up 193-fold from the 28 billion pkm in 1950 to 5.4 trillion pkm in 2012.[5] (See Figure 2.) The only pauses in an otherwise inexorable expansion came in 2001–02 (following the September 11 attacks in the United States) and in 2008–09 (after the start of the world financial and economic crisis).[6]

Like passenger air travel, air freight transport has expanded strongly. In 2012, some 49.2 million tons of goods were transported by plane worldwide.[7] Even though this is down 1 percent from 2011, it is 71 percent more than in 2001.[8]

The Asia-Pacific region, Europe, and North America dominate passenger and freight air transport, accounting for 84–86 percent of the world total, depending on the precise activity measured.[9] (See Table 1.)

International flights account for the bulk of air transport movements. In 2012, some 39 percent of all passengers were on board international flights, but because of the generally greater distances involved in such flights, cross-border trips accounted for 62 percent of all passenger kilometers.[10] In the same year, 66 percent

Michael Renner is a senior researcher at Worldwatch Institute and codirector of *State of the World 2014*.

Table 1. Regional Shares of Aircraft Departures, Air Passengers, and Air Transport, 2012

Region	Aircraft Departures	Passengers Carried	Passenger Kilometers	Freight Ton-Kilometers
		(percent)		
Asia-Pacific	25.3	31.2	30.2	32.3
Europe	24.7	27.0	27.2	26.3
North America	35.7	27.4	26.9	25.5
Middle East	3.4	4.9	8.2	9.1
Latin America and Caribbean	8.1	7.2	5.2	4.6
Africa	2.8	2.3	2.3	2.2
	(million)		(billion)	
World	31.2	2,957	5,402	687

Source: Based on data in ICAO, Annual Report of the Council 2012 *(Montreal: 2013).*

of freight tonnage was transported on international flights, which accounted for an even more imposing 86 percent of total freight ton-kilometers.[11]

According to the International Civil Aviation Organization (ICAO), the world's commercial air transport fleet grew from 18,972 planes in 2003 to 25,252 in 2012, an increase of 33 percent.[12] The largest fleet of aircraft by far is in the United States, which has about 6,000 planes in service, followed by China, with slightly less than 2,000.[13] All other countries have fewer than 1,000 planes in service each.[14] The bulk of the fleet is for passenger transport. The global airliner fleet is expected to grow considerably—reaching more than 36,500 planes by 2032 according to Airbus forecasts and more than 41,000 according to Boeing forecasts.[15]

Measured by revenue, 18 of the world's 50 largest airlines in 2010 were from the Asia-Pacific region, 12 were from Europe (including Russia), 11 from North America, 4 each from Latin America and the Middle East, and 1 from Africa.[16] The airline industry has long been marked by a series of mergers. Since the late 1990s, another phenomenon has been the creation of globe-spanning alliance groups—Star Alliance, SkyTeam, and oneworld.[17] (See Table 2.) These alliances account for almost 55 percent of all seats flown and 61 percent of available seat kilometers, with even higher shares for long-distance flights.[18] Companies that remain outside these alliances are mostly low-fare airlines and independent regional carriers.

Following many mergers and consolidations in the aircraft manufacturing industry, just two companies—U.S.-based Boeing and Europe's Airbus—account for the bulk of the world's airliner fleet. Boeing has delivered close to 11,000 planes in the last quarter-century, while its rival has sold close to 7,700.[19] (See Figure 3.) But Boeing has witnessed tremendous upswings and downswings, while

	Table 2. Airline Alliance Groups, 2010					
Group	Member Airlines	Countries	Airports	Number of Planes	Daily Departures	Million Passengers
Star Alliance	27	194	1,329	4,570	21,900	671
SkyTeam	18	187	1,000	4,300	> 15,000	537
oneworld	12	151	981	3,283	14,244	475

Source: Calculated on basis of data from Flightglobal, "World Airline Rankings: Regional Picture," 22 July 2011.

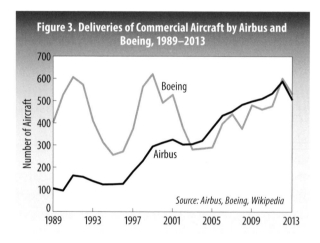

Figure 3. Deliveries of Commercial Aircraft by Airbus and Boeing, 1989–2013

Source: Airbus, Boeing, Wikipedia

Airbus has steadily gained, and during the last decade the two companies have been neck-and-neck competitors.[20]

In the market for smaller commercial jets (so-called narrow-body or single-aisle aircraft), the two companies will likely face rising competition from a number of firms such as United Aircraft Corporation (Russia), Commercial Aircraft Corporation (China), Bombardier (Canada), Embraer (Brazil), and Mitsubishi Heavy Industries (Japan).[21] A Boeing forecast suggests that the share of these smaller planes will rise from 64 percent of the global fleet in 2012 to 70 percent by 2032.[22]

In 2013, there were 41,821 airports in the world.[23] The United States had by far the largest number (13,513), followed at a distance by Brazil (4,093), countries in the European Union (3,102), Mexico (1,714), Canada (1,467), Russia (1,218), and Argentina (1,138).[24] A number of South American countries have several hundred airports each, as do Indonesia, South Africa, Papua New Guinea, China, and Australia.[25]

However, these raw numbers conceal numerous differences among airports. Many are very small in scale, while a select few have grown into the equivalent of small cities, occupying hundreds of square kilometers, processing millions of passengers each year, and employing tens of thousands of people.

Ranked by embarking or disembarking passengers, the world's top 25 airports together served 1.4 billion people in 2012, and they collectively accounted for 12 million of the world's 77 million aircraft movements (takeoffs and landings).[26] Atlanta had the most passengers (95.5 million), followed by Beijing, London (Heathrow), Tokyo, and Chicago (O'Hare).[27] Altogether, 10 U.S. airports are in the top-25 league, followed by 6 European, 4 Chinese, 4 other Asian, and 1 Middle Eastern.[28] This is a substantial change from earlier rankings. In 2000, in contrast, 17 of the top 25 airports (and 4 of the top 5) were in the United States.[29] The rankings change quite dramatically when only considering international passengers. In the 12 months ending August 2013, London (Heathrow), Dubai, Hong Kong, Paris (de Gaulle), and Amsterdam topped the listing, each with more than 50 million passengers.[30]

By cargo tonnage, Hong Kong (4.12 million tons), Memphis (4.05 million tons), Shanghai (2.97 million tons), Anchorage (2.47 million tons), and Seoul (2.46 million tons) were in the lead in 2012.[31] Both the scale of air shipments and the pecking order have changed over the years. In 2000, the largest cargo airports were Memphis (with 2.49 million tons), Hong Kong (2.27 million tons), Los Angeles (2.04 million tons), Tokyo (1.93 million tons), and Seoul (1.8 million tons).[32]

Aviation has a range of environmental and health impacts, including noise, land degradation, disturbance of wildlife and biodiversity, and emissions of air pollutants and greenhouse gases (GHGs). According to the International Council on Clean Transportation (ICCT), the cumulative climate impact of aviation to date is equivalent to about 40 percent of all surface transport modes, even though motor vehicles are far more numerous than planes.[33] Relative to all sources of GHG emissions, the sector is responsible for about 4 percent of climate change, and its role is rapidly rising.[34]

ICCT finds that design changes doubled the efficiency of commercial aircraft since 1960, but that progress has been slow in the last 20 years.[35] This has to do with low fuel prices for an extended period of time, as well as a tripling in the average age of aircraft since the late 1980s.[36] The higher fuel prices of more recent years provide an incentive to reinvigorate efficiency efforts. From levels below 70¢ per gallon throughout the 1990s, U.S. jet fuel prices rose in subsequent years and spiked at $3.89 in July 2008.[37] Following a brief crash in 2009, prices oscillated between $2.62 and $3.28 in the last three years.[38]

Still, fuel prices alone are an insufficient and unreliable driver. Government policy needs to provide a push for the development and use of more energy-efficient technologies, such as the global CO_2 standard for new aircraft under development at the ICAO.[39]

Broadly understood, airline fuel efficiency is the result of factors beyond technology, including a host of operational practices, load factors, seating density, and routing of flights (compared with the shortest possible travel distance).[40]

Thus, all of these dimensions need close scrutiny, and this calls for structural change as well as technical improvements. Beyond this, there are even more fundamental questions about whether the continued growth of the airline industry can be compatible with climate stabilization goals. It may be necessary to curtail both passenger and freight movements, with fundamental impacts on tourism, trade, and the broader economy.

Food and Agriculture Trends

Australian Department of Foreign Affairs and Trade

Gardener in Guguletu; Cape Town, South Africa

For additional food and agriculture trends, go to vitalsigns.worldwatch.org.

Agricultural Population Growth Marginal as Nonagricultural Population Soars

Sophie Wenzlau

The global agricultural population—defined as individuals dependent on agriculture, hunting, fishing, and forestry for their livelihood—accounted for 37.6 percent of the world's total population in 2011, the most recent year for which data are available.[1] This is a decrease of 12 percent from 1980, when the world's agricultural and nonagricultural populations were roughly the same size.[2] Although the agricultural population shrunk as a share of total population between 1980 and 2011, it grew numerically from 2.2 billion to 2.6 billion people during this period, principally in Africa and Asia.[3] (See Figure 1.)

Between 1980 and 2011, the nonagricultural population grew by a staggering 94.4 percent, from 2.2 billion to 4.4 billion people—a rate approximately five times greater than that of agricultural population growth.[4] In both cases, growth was driven by the massive increase in the world's total population, which more than doubled between 1961 and 2011, from 3.1 billion to 7 billion people.[5]

It should be noted that the distinction between these population groups is not the same as the rural-urban divide. Rural populations are not exclusively agricultural, nor are urban populations exclusively nonagricultural. For instance, the rural population of Africa in 2011 was 622.8 million while the agricultural population was 520.3 million.[6]

The statistics on the global agricultural population mask significant regional variation in trends, however. Although the agricultural population grew worldwide between 1980 and 2011, growth was restricted to Africa, Asia, and Oceania. During this period, this population group declined in North, Central, and South America, in the Caribbean, and in Europe.

For the last half-century, population growth in Africa—agricultural and otherwise—has been more pronounced than anywhere else in the world. Between 1980 and 2011, Africa's agricultural population grew by 63.2 percent and its nonagricultural population by 220.7 percent.[7] In 2011, the agricultural population of Africa accounted for 49.8 percent of its total population—a higher share than anywhere else in the world—and 19.9 percent of the world's total agricultural population.[8] (See Figure 2.)

Agricultural population growth during these years was second highest in Oceania, a region encompassing Australia, Melanesia, Micronesia, New Zealand, and Polynesia. Oceania's agricultural population grew by 48.5 percent, and its nonagricultural population grew by 65.5 percent.[9] Despite this growth, Oceania's agricultural population constituted only 0.3 percent of the world's total agricultural population in 2011.[10]

Between 1980 and 2011, the agricultural population of Asia grew by 20.5

Sophie Wenzlau is a senior fellow at Worldwatch Institute.

percent and the nonagricultural population by 134.4 percent.[11] Although both population groups grew less rapidly in Asia than in Africa, the total population of Asia (4.2 billion) remained four times that of Africa (1 billion) in 2011.[12] Due to the enormity of its total population, Asia's agricultural population constituted 74.5 percent of the world's total agricultural population and 46.4 percent of the people in Asia in 2011.[13]

The combination of movement to cities and agricultural consolidation caused agricultural populations to decline in Europe and the Americas between 1980 and 2011: by 66 percent in Europe, 44.6 percent in North America, 35.1 percent in South America, 13.6 percent in Central America, and 7.6 percent in the Caribbean.[14] (See Figure 3.)

Population trends have varied widely for the world's leading agricultural producers: China, India, and the United States. Between 1980 and 2011, the economically active agricultural populations of China and India grew by 33.2 and 50.7 percent, respectively, due to overall population growth.[15] The economically active agricultural population of the United States, on the other hand, declined by 37.2 percent as a result of large-scale mechanization, improved crop varieties, fertilizers, pesticides, and federal subsidies—all of which contributed to economies of scale (lower average costs of production as volume of output increased) and consolidation in American agriculture.[16]

Although we can only speculate about the reasons for growth and decline in various agricultural populations, it seems reasonable to conclude that overall population growth in Africa (256.2 percent), Asia (146.8 percent), and Oceania (130.9 percent) between 1961 and 2011 was the primary cause of agricultural population growth in these regions, particularly in their less-developed areas.[17] In regions where total population growth was lower, such as the Americas and Europe, agricultural population declined for the reasons just mentioned regarding U.S. agriculture.

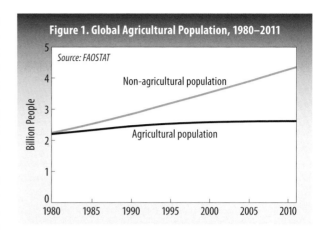

Figure 1. Global Agricultural Population, 1980–2011

Source: FAOSTAT

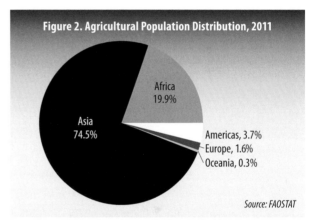

Figure 2. Agricultural Population Distribution, 2011

Source: FAOSTAT

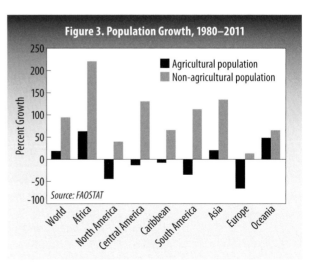

Figure 3. Population Growth, 1980–2011

Source: FAOSTAT

Although the world's agricultural population grew only marginally in recent decades, global agricultural output increased dramatically. According to the U.N. Food and Agriculture Organization (FAO), global net agricultural production increased by 112.1 percent between 1980 and 2011.[18] The world's net per capita production of agricultural goods increased by 34.9 percent during this period, averting food security crises in many places.[19]

At a global level, these remarkable production gains were driven, for the most part, by factors other than agricultural expansion. Between 1980 and 2011, the total area harvested of cereals, citrus fruit, oilcakes, pulses, roots and tubers, treenuts, and vegetables increased by only 16.8 percent, from 1 billion to 1.2 billion hectares, while production of these goods increased by 93.3 percent, from 2.6 billion to 5 billion tons.[20] (See Figure 4.) Agricultural yield (harvested production per unit of harvested area for crop products) of these items increased by 28.8 percent between 1961 and 1980 and by an additional 29.2 percent between 1980 and 2011, allowing farmers to feed a growing population without significantly expanding agricultural operations.[21]

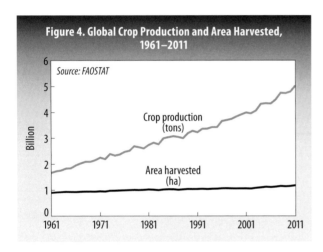

Figure 4. Global Crop Production and Area Harvested, 1961–2011

Source: FAOSTAT

This increase in agricultural productivity was fueled by the development of new and innovative technologies as well as by greater use of farm machinery, chemical fertilizers, pesticides, and irrigation systems. The numbers are telling: between 1975 and 2007, the world's net capital stock in agricultural machinery and equipment (in 2005 dollars) increased by 27.7 percent; between 1961 and 2011, the total area equipped for irrigation increased by 97.7 percent; and between 1961 and 2002, world fertilizer consumption increased by 353.1 percent.[22]

Although trend data are not available for world pesticide use, the U.S. Environmental Protection Agency estimates that 5.2 billion pounds of pesticides were used worldwide in both 2006 and 2007.[23]

According to the U.S. Department of Agriculture, productivity gains have been a driving force for growth in U.S. agriculture: between 1950 and 2000, the average amount of milk produced per cow increased from 5,314 pounds to 18,201 pounds per year, the average yield of corn rose from 39 bushels to 153 bushels per acre, and each farmer in 2000 produced on average 12 times as much farm output per hour worked as a farmer in 1950.[24] These gains can help explain the decline in U.S. agricultural population in the last 50 years.

Although productivity gains have enabled farmers to meet the growing demand for food, the methods used to achieve such gains have come with unintended consequences, including soil degradation, pollution, greenhouse gas emissions, and depleted freshwater supplies. Short-term production gains achieved by overusing chemical pesticides and fertilizers have, as a result, reduced the sector's long-term resilience to climate change.

The FAO estimates that the global agricultural population will decline by 0.7 percent and that the nonagricultural population will grow by 16.1 percent between 2011 and 2020.[25] The organization also estimates that feeding a population projected to reach 9.1 billion in 2050 will require raising overall food production by some 70 percent between 2005/07 and 2050.[26]

To address this challenge while promoting resilience to climate change and avoiding environmental degradation, farmers, governments, and the private sector could consider investing in agroecological approaches to farming—such as integrated pest management, no-till farming, cover cropping, and agroforestry—as well as programs designed to reduce food waste and food loss, which are currently estimated to be 1.3 billion tons per year (about one-third of all food produced for human consumption).[27] Policies encouraging the conversion of land from biofuels and livestock feed production to food production could also play a role in sustainably increasing the human food supply.

Global Food Prices Continue to Rise

Sophie Wenzlau

Continuing a decade-long increase, global food prices rose 2.7 percent in 2012, reaching levels not seen since the 1960s and 1970s but still well below the price spike of 1974.[1] Between 2000 and 2012, the World Bank global food price index increased 104.5 percent, at an average annual rate of 6.5 percent.[2] (See Figure 1.)

The price increases reverse a previous trend, when real prices of food commodities declined at an average annual rate of 0.6 percent from 1960 to 1999, approaching historic lows.[3] The sustained price decline can be attributed to farmers' success in keeping crop yields ahead of rising worldwide food demand. Although the global population grew by 3.8 billion (122.9 percent) between 1961 and 2010, net per capita food production increased by 49 percent over this period.[4] Advances in crop breeding and some expansion of agricultural land drove this rise in production, as farmers cultivated an additional 434 million hectares between 1961 and 2010.[5]

Food price volatility has increased dramatically since 2006. According to the U.N. Food and Agriculture Organization (FAO), the standard deviation—or measurement of variation from the average—for food prices between 1990 and 1999 was 7.7 index points, but it increased to 22.4 index points in the 2000–12 period.[6] Global cereal and dairy prices have shown significant volatility in recent years, whereas meat prices have fluctuated less.

Although food price volatility has increased in the last decade, it is not a new phenomenon. According to World Bank data, the standard deviation for food prices in 1960–99 was 11.9 index points higher than in 2000–12. Some price volatility is inherent in agricultural commodities markets, as they are strongly influenced by weather shocks.[7] But the recent upward trend in food prices and volatility can be traced to additional factors including climate change, policies promoting the use of biofuels, rising energy and fertilizer prices, poor harvests, national export restrictions, rising global food demand, and low food stocks.

International food price trends (measured in terms of consumer prices, not those paid to producers) varied by commodity in 2012. Due to the ubiquity of corn, wheat, and rice in global diets, changes in the price of cereal grains generally affect consumers more than fluctuations in other foods. Since food prices began increasing in the early 2000s, cereal prices have jumped more than 80 percent and exhibited significant volatility, according to FAO.[8] (See Figure 2.) Continuing this trend, global cereal prices increased 12.3 percent in 2012.[9] Unfavorable weather conditions—including severe drought in the United States and Eastern Europe—drove cereal prices up 18.2 percent between June and September, when they approached the all-time high observed in 2008.[10]

Despite an end-of-year decline in 2012, international wheat prices were 17

Sophie Wenzlau is a senior fellow at Worldwatch Institute.

percent higher in January 2013 than they were a year earlier, and maize prices were 11 percent higher.[11] The international price of rice moved up only marginally, by 0.4 percent.[12]

Domestic cereal prices exhibited significant variability from country to country in 2012. Prices increased most significantly in Malawi, Sudan, and Belarus and declined most significantly in Somalia, El Salvador, and Bolivia.[13] High cereal price volatility can have devastating consequences for the world's poor. The World Bank estimates that the 2011 food price spike—driven by a 57.9 percent increase in global cereal prices between June 2010 and April 2011—drove 44 million people into extreme poverty (under $1.25 a day).[14] High cereal prices can reduce real income and increase income instability, and they are associated with rising malnutrition as poorer people are forced to eat cheaper, less-nutritious food, as well as less overall. Significant declines in price, meanwhile, can increase poor farmers' vulnerability to bankruptcy.

Global meat prices increased by less than 1 percent between January 2012 and January 2013.[15] But annual averages can mask intermittent volatility, such as the 5.5 percent price surge between July 2012 and January 2013 that was triggered by higher grain—and thus animal feed—prices.[16] (See Figure 3.) The prices of feed-dependent poultry and pigmeat increased by 5 and 11 percent, respectively, between January and October, while beef and sheepmeat prices declined by 5 and 8 percent, respectively.[17] Growth in the global meat market is constrained by high feed prices as well as stagnating trade and consumption. In 2012, global meat production increased 1.6 percent to some 300 million tons, trade increased 2.2 percent, and consumption increased 0.4 percent.[18] These are all low rates compared with recent years.

In contrast to the relatively stable 1990–2005 period, international dairy prices have exhibited

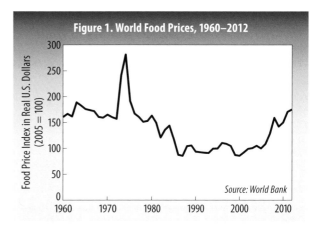

Figure 1. World Food Prices, 1960–2012

Source: World Bank

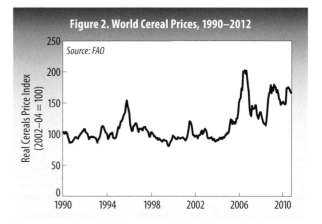

Figure 2. World Cereal Prices, 1990–2012

Source: FAO

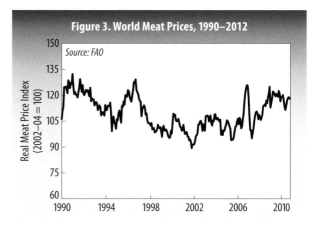

Figure 3. World Meat Prices, 1990–2012

Source: FAO

significant volatility since late 2006.[19] (See Figure 4.) The standard deviation of 14.3 index points during 1990–2005 more than doubled during 2006–13, to 30.5

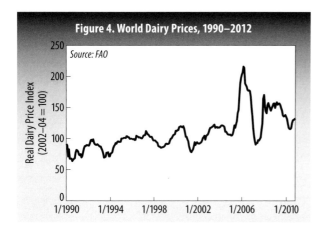

Figure 4. World Dairy Prices, 1990–2012

Source: FAO

index points.[20] Between January 2012 and January 2013, global dairy prices decreased by 4.2 percent.[21] According to the FAO, total milk production increased 3 percent in 2012 due to gains in Asia, Oceania, and South America; total dairy trade increased 4.6 percent, and per capita dairy consumption increased 1.8 percent.[22] The international market remains sensitive to sudden changes in milk production because of low government stores in the European Union and United States.

Various forces affecting global food supply and demand have influenced the level and volatility of food prices in the last decade. Population growth and increasing affluence—predominantly in Asia—have led to rising food demand since 2000, which in turn has triggered higher global food prices. Between 2000 and 2010, Asia's population grew 12 percent, from 3.7 billion to 4.2 billion, and in 2010, Asians accounted for 60 percent of the world's population.[23] In South-Central and Southern Asia alone, the population increased by 16.5 percent and 16.7 percent, respectively. Meanwhile, wages nearly doubled in Asia from 2000 to 2011, whereas they increased only 18 percent in Africa and 15 percent in Latin America and the Caribbean.[24]

As the global population and wages increased, growth in the world food supply declined from 0.44 percent annually on average during 1990–99 to 0.35 percent during 1999–2009.[25] Reduced agricultural research and development by governments and international institutions may have contributed to the slower growth in crop yields.[26] But the global food supply also has been affected by weather shocks, poor harvests, climate change, and a decline in agricultural total factor productivity (a measure of the efficiency with which inputs such as chemicals, labor, and machinery are transformed into crop yields). Perhaps most significant has been an increase in biofuels production in the last decade.

Between 2000 and 2011, global biofuels production increased more than 500 percent, due in part to higher oil prices and the adoption of biofuel mandates in the United States and the European Union (EU).[27] The EU, for example, has mandated that biofuels account for 10 percent of transportation fuel use by 2020.[28] Rising demand for ethanol and biodiesel has strained the global food system, leading to the diversion of food-producing farmland to produce biofuel feedstock, predominantly in Argentina, Brazil, Canada, China, the EU, and the United States.[29] According to a 2012 study by the University of Bonn's Center for Development Research, if biofuel production continues to expand according to current plans, the price of feedstock crops (particularly maize, oilseed crops, and sugarcane) will increase more than 11 percent by 2020.[30]

Other factors that affected global food prices in the last decade included higher-than-normal imports, trade policies, low levels of stocks, rising energy and fertilizer prices, and increased trade within futures markets for food commodities.

Large-scale imports of agricultural commodities in 2007–08 and 2011 were important factors in the global food price spikes in those years. High Chinese imports of soybeans, for instance, contributed to the 2011 spike.[31] National export restrictions, including taxes and bans, also drove up food prices; policies enacted in 2007–08 in response to the price spike generated panic in net-food-importing countries and raised grain prices by as much as 30 percent, according to some estimates.[32] In a 2011 FAO sample of 105 countries, 31 percent were found to have adopted one or more food export restrictions between 2007 and the end of March 2011.[33] Fifty percent of Asian countries applied restrictive export measures, compared with 21 percent in Africa and 18 percent in Latin America and the Caribbean.[34]

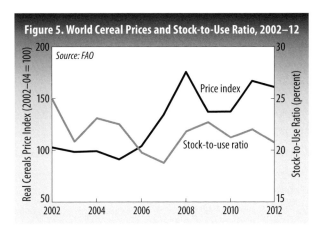

Figure 5. World Cereal Prices and Stock-to-Use Ratio, 2002–12

Source: FAO

In the last few decades, periods in which the cereals stock-to-use ratio (the level of carryover reserves of cereals as a percentage of total annual use) was near its minimum have correlated with a high price of calories from food commodities. When food stocks are high, shocks can be absorbed more easily than when stocks are low or nonexistent.[35] The world stock-to-use ratio for calories from wheat, maize, and rice was lower in the last decade than in the two preceding decades, which may be a main reason for higher global food prices.[36] Low aggregate stocks in 2007–08—which contributed to that period's global food price spike—can be attributed to both the diversion of maize to biofuels and unprecedented income surges in China and India, which generated a substantial increase in the demand for meat and maize-based feed.[37] (See Figure 5.)

Rising energy and fertilizer prices drove up food prices as well, by adding to production, processing, transportation, and storage costs. According to the World Bank commodity price index, the average price of energy during 2000–12 was 183.6 percent higher than the average price during 1990–99, while the average price of fertilizer increased 104.8 percent in the same period.[38]

Food producers and buyers have gambled on the futures market for agricultural commodities for centuries, using the market as a form of insurance against unanticipated price changes. Although the volume of speculation on food commodities futures has increased in the last decade, particularly in the United States since the repeal of the Glass-Steagall Act in 1999, the extent to which speculation has influenced food prices remains uncertain. In February 2012, key Eurozone banks, including BNP Paribas, Crédit Agricole, and Barclays, pulled back from speculative trading in agricultural commodities in response to criticism from Oxfam and other civil society groups that they were "profiting from hunger."[39] More research is needed to understand the role of speculation in agricultural commodities.

There is reason to believe that food commodity prices will be both higher and more volatile in the decades to come. As climate change increases the incidence of extreme weather events, production shocks will become more frequent; rising temperatures will also likely affect the average global food supply by reducing production capacity. Food prices will also likely be driven up by population growth, increasing global affluence, stronger linkages between agriculture and energy markets, and natural resource constraints. According to the FAO, although high food prices tend to aggravate poverty, food insecurity, and malnutrition, they also represent an opportunity to catalyze long-term investment in agriculture, which could boost resilience to climate change and augment global food security.

Agricultural Subsidies Remain a Staple in the Industrial World

Grant Potter

In 2012, the most recent year with data, agricultural subsidies totaled an estimated $486 billion in the top 21 food-producing countries in the world.[1] These countries—the members of the Organisation for Economic Co-operation and Development (OECD) and seven other countries (Brazil, China, Indonesia, Kazakhstan, Russia, South Africa, and Ukraine)—are responsible for almost 80 percent of global agricultural value added in the world.[2] OECD countries alone spent $258.6 billion in subsidies to support farming in their respective countries in 2012.[3] OECD subsidies grew rapidly between 2001 and 2004, rising from $216 billion to over $280 billion.[4] Since then, the dollar amount received by OECD farmers has stayed roughly static at between $240 billion and $280 billion.[5] (See Figure 1.) But from 2001 to 2012, the amount spent on these subsidies as a percentage of the total value of agriculture produced in the OECD declined steadily from 32 percent to 19 percent.[6] This means that for every dollar's worth of agriculture earned by OECD farms in 2012, 19¢ came from some kind of government subsidy policy.[7]

Agricultural subsidies are not equally distributed around the globe. In fact, Asia spends more than the rest of the world combined.[8] (See Figure 2.) China pays farmers an unparalleled $165 billion.[9] Significant subsidies are also provided by Japan ($65 billion), Indonesia ($28 billion), and South Korea ($20 billion).[10] Europe also contributes a great deal to agricultural subsidies due in large part to the Common Agricultural Policy (CAP) of the European Union (EU). At over $50 billion, CAP subsidies accounted for roughly 44 percent of the entire budget of the EU in 2011.[11] And this figure does not even include EU price supports, in which governments keep domestic crop prices artificially high to give farmers a further incentive at the expense of the consumer. Including these price supports, the EU spent over $106 billion on agricultural subsidies in total.[12] North America provides almost $45 billion in subsidies, with the United States spending just over $30 billion and Canada and Mexico spending $7.5 billion and $7 billion, respectively.[13] Of the countries studied by the OECD, 94 percent of subsidies were spent by Asia, Europe, and North America—leaving only 6 percent for the rest of the world.[14]

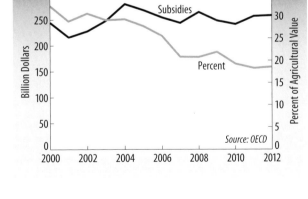

Figure 1. OECD Agricultural Subsidies and Share of Value of Agriculture Produced, 2000–12

Source: OECD

Grant Potter was a development associate at Worldwatch Institute.

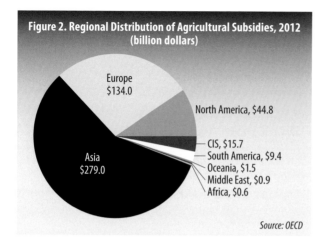

Figure 2. Regional Distribution of Agricultural Subsidies, 2012 (billion dollars)

Europe $134.0

North America, $44.8

CIS, $15.7
South America, $9.4
Oceania, $1.5
Middle East, $0.9
Africa, $0.6

Asia $279.0

Source: OECD

The term "subsidies" covers a vast number of different policy options, but at the heart of all of them is government intervention in agricultural markets. A common type of subsidy is called direct payments. These are regularly paid to farmers who produce a designated crop (in the United States, until recently the crops were wheat, corn, sorghum, barley, oats, cotton, rice, soybeans, minor oilseeds, and peanuts), and the payments are decoupled from production—which means that farmers can produce as much or as little as they want and still receive this subsidy.[15] Direct payments are the cornerstone of the EU CAP and account for $40 billion of its $50-billion budget.[16]

Direct payments were a staple of U.S. farm policy from 1996 to 2013, but the $5 billion spent on direct payments was struck from the U.S. Farm Bill signed by President Obama in February 2014.[17] In lieu of these payments, the United States expanded the federally subsidized crop insurance program to $9 billion a year.[18] This allows farmers to buy crop insurance premiums, 60 percent of which is subsidized by the government; private insurance companies compensate farmers for lost revenue if crop prices drop too low or if the farmers have low yields in a given year.[19] During the drought of 2012, insured American farmers received $16 billion.[20] Many people have criticized crop insurance as wasteful because farmers can often earn more money from insurance in a bad year than they can from selling their crops in a good year. A prime example of this is corn and soybean farmers, who in 2012 earned $12.7 billion when their actual economic losses only amounted to around $6 billion.[21] Bruce Babcock, a professor of economics at Iowa State University, predicts that insurance will cause overproduction as farmers will overplant and manage crops less carefully in order to reap more insurance payouts in bad years.[22]

Subsidies like direct payments and crop insurance are criticized as not being safety nets for poor farmers, as is their stated purpose, but rather a way for wealthy farmers to get richer. The direct payment policy of the EU CAP, called the Single Farm Payment, is distributed by the hectare—so farmers who own or rent more land receive greater financial benefits.[23] The BBC reported that in the United Kingdom in 2012, "889 landowners received more than £250,000. Of those, 133 were given more than £500,000 and 47 of those were given more than £1m in subsidy."[24] In the United States, the newly expanded crop insurance program receives similar criticism. The Environmental Working Group estimates that in 2011, more than 10,000 farms received between $100,000 and $1 million in federal crop insurance subsidies, and 26 farms received more than $1 million.[25] In contrast, the bottom 80 percent of farms (389,494 holdings) individually received only $5,000 that year.[26]

By predominantly funding a few staple crops for the largest farms, subsidies support industrial-scale operations. These factory farms tend to lack crop diversity, which over time saps the soil of nutrients and in turn requires substantial use of artificial chemical inputs like fertilizers and pesticides.[27] One example of this is in the U.S. Midwest, where farms rotate between corn and soy, which requires a significant use of synthetic nitrogen fertilizer to achieve adequate yields. One side effect of this is that fertilizer runoff into the Gulf of Mexico is responsible for a massive algae bloom. This area, which grew alarmingly from 2,900 square miles in 2012 to 5,840 square miles in 2013, has been called a "dead zone" because the algae suck oxygen out of the water and prevent other aquatic life from developing.[28]

Price supports are another category of subsidies. They are intended to keep domestic crop prices high enough to encourage farmers, even during periods of overproduction when prices would tend to fall due to oversupply. This can be achieved by imposing a tariff or quota on agricultural imports so that potentially cheaper foreign agricultural products cannot drive down domestic prices.[29] In conjunction with import barriers, a price-supporting government can offer to buy agricultural commodities at a certain price; this raises the price to consumers, as farmers will refuse to sell below the government's support price.[30] Price supports make up almost 70 percent of China's subsidy spending, and the government particularly encourages the growth of staple crops such as rice and wheat.[31] China increases the minimum price for rice and wheat yearly; between 2007 and 2012 the price doubled for rice and increased by 70 percent for wheat.[32] If the price for either wheat or rice falls below this minimum, the state-owned China Grain Reserves Corporation will continue to make purchases at the minimum price until what farmers can ask for rises above that number.[33]

Price supports can cause overproduction and oversupply because they encourage greater production of a specific commodity and less domestic consumption (since consumers tend to buy less as prices rise).[34] Rather than let the oversupply go to waste, it is traded on the international market at artificially low prices. Since subsidized farmers are insulated from the true cost of farming, they can afford to sell at a lower price than their less-subsidized foreign counterparts.[35] Many developing countries have argued that this undermines their own agricultural sectors as they cannot afford to spend billions in subsidies to overcome this handicap.[36]

This disparity is particularly obvious when farmers in industrial countries are pitted against farmers from the least-developed countries. Due to the importance of agriculture in their economies and the relatively lower cost of labor, these countries have a comparative advantage in farming over the industrial world.[37] However, the U.N. Conference on Trade and Development (UNCTAD) argues that this advantage evaporates in the face of heavy subsidization and cites a number of cases where the least-developed countries opened their markets to low-priced subsidized crops that ended up crippling their local agricultural sector.[38] According to UNCTAD, these countries turned rapidly from net exporters to net importers, and between 2002 and 2008, they saw their food imports jump from $9 billion to $24 billion.[39] The inability of developing countries to compete against subsidy-backed crops can be seen in a trade deal between the United States and sub-Saharan Africa called the African Growth and Opportunity Act.[40] The Brookings Institution estimates that agriculture accounted for less than 1 percent of exports from sub-Saharan Africa to the United States as part of this deal even though two-thirds of the population there is involved in agriculture.[41]

Agriculture is one of the most contentious subjects at World Trade Organization (WTO) negotiations due to the tension over subsidies. In Geneva in 2006, the Doha Development Round fell through in part because the United States was unwilling to reduce trade-distorting subsidies further and because developing countries were unwilling to reduce import tariffs for U.S. crops further.[42] Following a long period of stalled negotiations, a Bali Package was finally agreed to in December 2013, which the WTO considers a major source of hope for a more harmonious future for agriculture and trade. The Bali Package includes, among other things, a pledge to reduce export subsidies and agreements on how to administer agricultural tariffs.[43]

Global Economy and Resources Trends

Brick factory in the Mekong Delta of Vietnam

For additional global economy and resources trends, go to vitalsigns.worldwatch.org.

Global Economy: Looks Good from Afar But Is Far from Good

Mark Konold and Ralph Albus

Gross world product increased to just over $83 trillion in 2012, a 4.85 percent increase over 2011.[1] (See Figure 1.) On the surface, this metric supports the argument that the worst of the global recession is in the past—and it is. However, closer inspection shows that while there is at least tepid growth globally, growth rates like the ones experienced in the prior 20–25 years now seem to be taking place in emerging economies. In fact, the total gross domestic product (GDP) of emerging economies is now roughly equal to that of all the advanced economies.[2]

Nominally, a growth rate just shy of 5 percent seems reasonable, but this rate of growth continues a pattern of slowing growth rates since 2010 and 2011, which were 6.35 and 5.67 percent, respectively.[3] Ironically, one trend that seems continuous since 1980 is that the percent change in global GDP varies widely. (See Figure 2.)

The gross world product is the sum of the GDPs of all countries. This typically includes levels of consumption, investment, government spending, the cost of imports, and the proceeds from exports.[4] Because of various transaction costs, floating exchange rates, and barriers such as tariffs, a metric is applied to put purchasing power for countries on an even footing. This factor, applied to the figures in this article, is called the purchasing power parity exchange rate.[5]

Economic activity continued to be a muddled picture in 2012, with some parts of the globe seeing relatively strong growth while others had tepid growth and some actually experienced some negative growth. In recent years, developing Asia has had the highest growth rate of any region by far.[6] (See Figure 3.)

Indeed, the strongest areas of growth continued to be developing and emerging economies due to policies that created favorable business and investment climates.[7] Countries that avoided overleveraging themselves while experiencing robust growth in recent years have had relatively healthier fiscal positions and therefore higher levels of foreign direct investment.[8] However, this growth was slightly hindered as advanced economies like the United States, Japan, and the members of the European Union (EU) dealt with headline-grabbing crises such as the "fiscal cliff" and the possible breakup of the EU. Despite successfully dealing with these challenges in 2012, there was little progress on long-term solutions, which may have kept growth modest.[9]

Other metrics show that the global economy sputtered a bit in 2012. Trade flows, often a bellwether of economic health, continued to slow. As a result, current account balances, a collection of all transactions other than financial and capital, continued narrowing throughout the year, down from $355 billion in 2011 to $343 billion in 2012, a 3.4 percent drop.[10] Equally noteworthy was the down-shift in commodity prices in 2012, which was significant enough to mark the end

Mark Konold is the Caribbean program manager for Worldwatch's Climate and Energy Program. Ralph Albus was a Climate and Energy research intern at the Institute.

of a cycle that saw overly robust growth in the preceding 10 years.[11] Non-energy-related commodities were down by 7.5 percent, while oil prices stayed relatively flat throughout 2012.[12]

Unemployment levels also indicate that, despite continued growth, economic health is far from a rosy picture. According to the International Labour Organization (ILO), 200 million people around the world are unemployed—about 6 percent of the global workforce.[13]

In addition, the ILO notes that the labor share of national income has been steadily falling around the world for some time, with a larger share of the income going to capital because of the expansion of technological innovation, decreased labor union participation, globalized trade, and wider and more-interconnected financial markets.[14] The ILO estimates that between 1999 and 2011 (the last year with complete data), labor productivity in advanced economies increased twice as much as average wages.[15] ILO data show how wage growth has sputtered in recent years. Excluding China, real wage growth has averaged less than 1 percent a year since the onset of the financial crisis.[16] (See Figure 4.) Low rates of wage growth may be contributing to slow economic growth given the positive correlation between labor compensation and household consumption.[17]

While global GDP gives a broad-stroke picture of economic health, it is incomplete. It makes no distinction between the money spent building a hospital, a school, or a prison. It does not differentiate between sales of gasoline, timber, or alcohol, nor is there any distinction between money spent on health care, education, or disaster cleanup. In addition, it does not consider things like home production, the value of time spent keeping up a home and rearing children, or the return on consumer durable goods.[18] According to one study, while the factors and calculations are complex, adding them to the measure of gross national product (GNP) would have resulted in a value 25 percent higher than actually recorded.[19] (GDP became the primary measure of growth in 1991, while prior to that, GNP was the accepted metric. The former is a measure

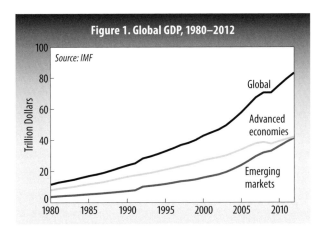

Figure 1. Global GDP, 1980–2012

Source: IMF

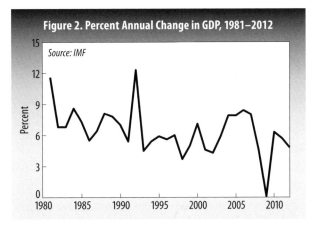

Figure 2. Percent Annual Change in GDP, 1981–2012

Source: IMF

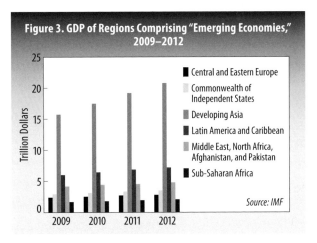

Figure 3. GDP of Regions Comprising "Emerging Economies," 2009–2012

Central and Eastern Europe
Commonwealth of Independent States
Developing Asia
Latin America and Caribbean
Middle East, North Africa, Afghanistan, and Pakistan
Sub-Saharan Africa

Source: IMF

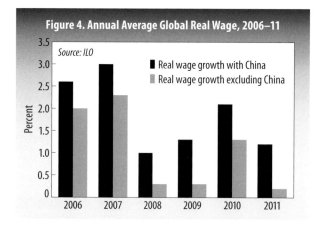

Figure 4. Annual Average Global Real Wage, 2006–11

Source: ILO

■ Real wage growth with China
■ Real wage growth excluding China

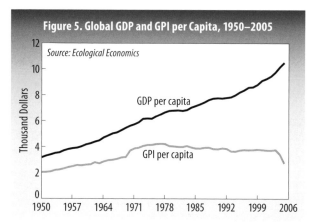

Figure 5. Global GDP and GPI per Capita, 1950–2005

Source: Ecological Economics

GDP per capita

GPI per capita

of total income produced domestically. The latter includes only income earned by nationals of a country. The study cited contained data prior to 1991 and therefore cited GNP throughout.)

Conventional economics regards economic growth as an unalloyed good, necessary to improve human well-being. But it is only a nominal indicator. It lacks the many intricacies and more subjective goods that are essential to a more encompassing, holistic, meaningful metric. Given the disparity of the benefits, it is clear that in and of itself, growth is far from an effective measuring stick.

In light of this, new metrics have been developed that seek to paint a more comprehensive and accurate picture of humanity's overall welfare. An enhanced and widely cited metric is the Genuine Progress Indicator (GPI), which includes the economic cost of expenditures that diminish "community capital."[20] Though GPI has critics for what it includes and excludes, it is more comprehensive than GDP and thus gives a more adequate picture of well-being. However, some critics argue that it is still rooted in a consumption-based model and ignores the important issue of sustainability or preservation of resources.

Still, the study of GPI has yielded some interesting results. For example, a 2012 study looked at a group of countries that contain more than half of global population and for which data were either tracked or could be reasonably estimated.[21] It showed that after the mid-1970s, per capita figures for GDP and GPI, which had shown a strong correlation up to that point, began to diverge. While GDP per capita continued to rise, broader economic well-being leveled off and has even declined.[22] (See Figure 5.)

Other attempts to gauge human progress have emerged recently. In 2011, taking a cue from Bhutan, which began tracking something called Gross National Happiness in the 1970s, the United Nations General Assembly passed a resolution stating that all countries should begin measuring happiness.[23] And in 2013, the second World Happiness Report, sponsored by the Sustainable Development Solutions Network, was released to provide some insight for discussions about a global policy for sustainable development.[24]

The new report tracks and analyzes such factors as social support, freedom to make life choices, perception of corruption, and healthy life expectancy at birth.[25] It demonstrates that despite the financial crisis in 2007–08, overall happiness has been on the rise.[26] However, this trend has not been universal. Increases in happiness have mostly been recorded in sub-Saharan Africa, East Asia, the Commonwealth of Independent States, Latin America, and the Caribbean.[27] Larger western economies, particularly the United States and the countries affected by the Eurozone crisis, have seen happiness drop.[28] There have also been declines in the Middle East, North Africa, and South Asia.[29]

Similarly, the U.N. Development Programme prepares a Human Development Index as an indicator of well-being that relies primarily on health (life expectancy at birth), education (mean and expected years of schooling), and living standards (gross national income per capita on a logarithmic scale.)[30]

Regardless of the approach or specific metrics involved, all studies seem to conclude that sole reliance on GDP growth as the measuring stick for prosperity and well-being is woefully inadequate and probably has been for some time. Higher human development can only begin to be seriously measured when the impact of expenditures is evaluated and when more-qualitative elements are examined as part of a much larger and more complex mechanism to decipher human development.

But even when these more-qualitative factors are considered, most approaches to global economic health ignore the planet's capacity to provide the resources necessary to sustain it. The resources of the natural environment are finite, while human consumption has reached such a point that humanity consumes a year's worth of resources in less than 365 days. Groups like the Global Footprint Network strive to accurately account for resource use by measuring humanity's impact on the ecosystem compared with the ecosystem's ability to replenish itself.[31] The organization tries to determine the rate at which humanity's use of natural resources exceeds the resources Earth can sustain. In fact, the group widely publicizes its annual Overshoot Day. The first time the event was noted was in 1987, on December 21st. By 2012, Overshoot Day had crept back to August 22nd.[32] Current studies note that civilization is above Earth's capacity to replenish itself by about 50 percent.[33]

More Businesses Pursue Triple Bottom Line for a Sustainable Economy

Colleen Cordes

As corporations of all sizes increasingly choose to monitor and report on their social and environmental impacts, a growing number of mostly small and medium-sized companies are going even further: they are volunteering to be held publicly accountable to a new triple bottom line—prioritizing people and the planet as well as profits.[1]

Just how broadly, rapidly, and rigorously this movement can spread is of critical importance, given the supersized global impacts of for-profit enterprises. Sustainable economies are likely to remain elusive without substantial shifts in corporate norms. Recent data provide signs that such change is possible and indeed may even have begun.

Over the last 15 years, for example, the number of businesses of all sizes that choose to self-assess how sustainable their operations are—using widely accepted social and environmental standards—and to publicly disclose their results has been growing rapidly, especially in Europe and Asia.[2] (See Figure 1.)

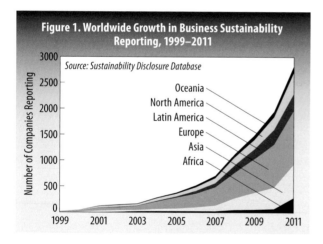

Figure 1. Worldwide Growth in Business Sustainability Reporting, 1999–2011

Source: Sustainability Disclosure Database

And within this expanding universe, two dramatic trends are developing among some companies, most of which are small or medium-sized.[3] They are seeking either a new legal requirement or a third-party certification that will hold them accountable to higher company-wide standards, in terms of trying to achieve positive social and environmental impacts in addition to the conventional corporate goal of earning a profit. In effect, then, they are embracing a triple bottom line.

The more recent of these developments is the rise of a fast-moving movement, with significant leadership provided by sustainably minded businesses.[4] Their goal is to persuade lawmakers to create a new legal status known as "benefit corporation" that for-profit businesses can choose voluntarily.

A "benefit corporation" is a corporate form that requires a company to legally establish in its original or amended articles of incorporation that it has a general purpose of having a positive impact on society and the environment and that its board of directors, in making decisions, is required to take into account the interests of multiple stakeholders in addition to the financial interests of its shareholders. The stakeholders it must consider, by law, include the company's own

Colleen Cordes is a public policy consultant and director of outreach and development at The Nature Institute of Ghent, New York.

workforce and that of its suppliers, its customers, the local community and general society, and the local and global environment.[5]

Benefit corporations are also required to report annually and publicly on their overall social and environmental impact as assessed against a transparent, credible, and independent third-party standard.[6] (The laws generally do not state that the third party itself must actually perform or certify the assessment, although at this point, there is some confusion about that aspect.)

Proponents of this new corporate form say it essentially bakes a triple bottom line into a company's DNA. That frees companies from the fear of shareholder law-suits if their decisions fail to maximize shareholder value because of some compet-ing interest of other stakeholders, such as workers. Under current corporate case law in the United States, for example, corporate directors are generally assumed to be liable in such suits. Incorporation as a benefit corporation is intended to establish the directors' fiduciary responsibility to consider the interests of all stake-holders. Formalizing a company's social and environmental purposes under a legal framework also makes it more likely that its good intentions will survive the depar-ture of its founders or any major spurts of growth and that its directors will have the legal backbone to fend off buyout offers from conventional corporations that do not have the same commitments.[7]

The legal sufficiency of benefit corporation status has not yet been tested in the courts, however, as the statutes have been enacted so recently. Also, some nonprofits are concerned that benefit corporations may seek special tax treat-ment or other exceptional treatment from governments at the expense of the nonprofit community.[8]

The movement for benefit corporation statutes began in the United States, under the leadership of a U.S. nonprofit, B Lab, which developed model legislation with the pro bono help of U.S. law firms.[9] Maryland and Vermont became the first states to pass such legislation, which they did with strong bipartisan support.[10]

The benefit corporation movement is still primarily a phenomenon in the United States: 13 states have now signed such legislation into law, and similar legislative proposals have been introduced in another 15 states and the District of Columbia. (See Table 1.) There is significant interest in such legislation in several other states, as well.[11] Because some states do not keep records on how many com-panies elect to become benefit corporations, it is impossible to report the current total accurately. However, B Lab reports that by early April 2013, about 70 of the companies that it had separately approved as Certified B Corporations have also officially registered as benefit corporations in their home states.[12] B Lab, which also compiles whatever information states do make available, estimates that there are currently about 200 benefit corporations in the United States (none of which are publicly traded companies at this point).[13]

Most benefit corporations to date are either small or medium-sized busi-nesses.[14] But they include a few larger companies that are privately held, such as the outdoor apparel and accessory firm Patagonia Inc., which reportedly had annual sales of about $540 million for the year ending April 2012, and King Arthur Flour, an employee-owned, 223-year-old company with reported sales of about $84 million in 2010.[15]

Table 1. U.S. Movement for Benefit Corporation Laws	
State Statute and Month Effective	**Legislation Introduced and Status (as of mid-April 2013)**
Arkansas, signed into law April 2013*	Alabama, passed Senate committee
California, January 2012	Arizona, passed Senate and House
Hawaii, July 2011	Colorado, passed Senate and House
Illinois, January 2013	Connecticut
Louisiana, August 2012	Delaware
Maryland, October 2010	District of Columbia, waiting congressional approval
Massachusetts, December 2012	Florida
New Jersey, March 2011	Iowa
New York, February 2012	Montana, passed House
Pennsylvania, January 2013	Nevada, passed Assembly
South Carolina, June 2012	New Mexico, passed both houses; Governor pocket-vetoed in early April
Vermont, July 2011	North Carolina
Virginia, July 2011	Oregon, passed House committee
	Rhode Island
	Texas
	West Virginia

*Effective 90 days after end of legislative session.
Source: Benefit Corp Information Center, "State by State Legislative Status," at www.benefitcorp.net.

Outside the United States, B Lab has partnered with an organization in Chile, Sistema B, to expand the movement to South America. Sistema B has worked in Chile, Argentina, Colombia, and Brazil, exploring with local partners whether the legal infrastructure is in place to allow companies to write or amend their articles of incorporation and bylaws to stipulate that they require themselves to consider the interests of all their stakeholders in making decisions. B Lab reports that a national legislative proposal to provide that option was likely to be introduced in spring 2013 in Chile.[16]

B Lab also reports that it is exploring additional regional partnerships to help expand its scope. That includes assistance in researching and developing, where needed, proposals to provide the legal infrastructure to protect companies seeking to establish a fiduciary responsibility for directors and officers to consider the interests of a broad range of stakeholders. There is already some interest in Europe, and research has begun on the corporate code in the United Kingdom. More generally, companies in 25 countries outside the United States have earned B Lab's third-party certification as Certified B Corporations.[17] Canada and Chile are the two countries with the most activity outside the United States.[18] As the number in any country begins to grow, B Lab plans to work with companies that are interested in exploring the need and opportunities for revisions in their home countries' legal infrastructure. A few companies in Australia have already expressed an interest in this.[19]

In effect, the community of Certified B Corporations, or B Corps, as they are informally called, is a growing advocacy group for the benefit corporation movement. Before a B Corp is recertified or certified for the first time, B Lab requires the company to have begun working toward establishing legally in its governing documents that the board and officers must consider the interests of all stakeholders—not just shareholders—when making decisions if the company is headquartered in a location where such a requirement is now legally available. This could be done either by filing for benefit corporation status or by amending articles of incorporation under a constituency statute, if one exists.

In the United States, for example, the corporate codes in 30 of the 50 states do include constituency statutes, which allow but do not require companies to consider the interests of other stakeholders besides shareholders in making decisions.[20] But there is little case law to determine whether directors and officers who make decisions based on the interests of other stakeholders, such as workers or the local community, would be held liable if such decisions failed to maximize profits.

Where neither legal option exists, Certified B Corps comprise a natural constituency to advocate for such change, increasing the potential for country-by-country advocacy for this new kind of corporation with a general purpose of positive social and environmental impact.[21]

Other major organizations are now partnering with B Lab to promote benefit corporation legislation in the United States. The new American Sustainable Business Council represents more than 62 business associations, which in turn represent more than 165,000 businesses.[22] It has established promotion of the benefit corporation movement as a major public policy goal. Green America is a U.S. nonprofit that created the first U.S. green business network in 1983, at a time when "green" and "business" were two words not commonly found together.[23] The organization, which offers a certification of companies' social and environmental practices that is focused on small to mid-size businesses, also has supported the movement.[24]

A second notable trend, in terms of businesses aspiring to hold themselves to higher standards of overall social and environmental impact, is the growing number of for-profit businesses that are seeking and winning company-wide certification from third-party organizations. (Evaluating and comparing the rigor of third-party certifications is beyond the scope of this article.)

Again, as with benefit corporations, most for-profits seeking such approvals to date are small to medium in size, although some companies with revenues in the hundreds of millions, such as Patagonia and Seventh Generation, are also included here.[25] Initially, such certifications were most often sought and won by small enterprises, often very small, which could be less constrained by outside investors' interest in maximizing profits. Over time, however, as the sustainability movement has expanded and a much broader range of companies understand the advantages of public recognition for corporate social responsibility, larger companies are expressing interest as well.[26]

Third parties, such as B Lab or Green America, that officially evaluate whether for-profit enterprises as a whole meet a required standard base their award on detailed assessments across environmental, social, and governance (ESG) criteria. (There are now hundreds of certifications that businesses can apply for, but the vast majority are applied to particular products or services or focus on a limited range of indicators, such as fair trade or environmental impact, often limited to a particular sector of the economy. B Lab and Green America are among the relatively few organizations that provide broad, company-wide evaluations and approvals.)[27]

The Green Business Network (GBN) at Green America (formerly the Co-op America Business Network) started in 1983 with 363 approved companies.[28] By 1993, the number of approved companies had grown to 1,300; by 2003, it was up to 2,000, and by early 2013, it was about 3,500.[29] About 50 GBN members are headquartered outside of the United States, mostly in Canada.[30]

In late 2011, Green America shifted to a new certification process, for the first time recognizing different levels of sustainability performance to match what the organization described as an "explosive growth" in the number of businesses trying to improve their sustainability performance. Only the highest of three levels, Gold, is based on an independent evaluation by Green America, and the criteria for it are similar to those that companies formerly had to meet to be approved as GBN members. Membership at the Bronze and Silver levels is now awarded based on a company's self-reports, using Green America's online assessment tool.[31]

Green America introduced the first screening standards in the United States, and perhaps in the world, that assessed businesses' "familiarity with, commitment to, and action toward social and environmental responsibility." Its methodology includes more than 800 sustainability actions on which Green America assesses businesses.[32] Since 1983, it has screened and provided individual assistance to more than 8,000 businesses.[33]

Companies that have won B Lab's certification as B Corporations have grown from 78 in 2007, the first year such certification was available, to 725 by early April 2013.[34] (See Table 2.) Total gross revenues for all Certified B Corps are about $6 billion annually, and together these businesses employ about 30,000 people, according to B Lab.[35]

Companies are also able to log in and confidentially fill out B Lab's online assessment for free, without applying to be officially evaluated for certification. Many companies do this, according to B Lab staff, as a benchmark to evaluate their performance and to identify improvements they need to make in order to score well enough to be certified. The number of companies annually using B Lab's online assessment tool, a marker for broader interest in eventual certification, grew from 280 in 2007 to 2,406 in 2012.[36] By the end of the first quarter of 2013, some 8,000 individual companies had used the tool.[37]

The year-to-year trends reported in the Sustainability Disclosure Database of the Global Reporting Initiative (GRI) set the two trends just discussed into a larger context: the increasing number of business enterprises worldwide that are at least self-assessing their social and environmental impact, using measures that are broadly recognized and accepted, and disclosing their results. The most commonly used measure is GRI's Sustainability Reporting Guidelines, developed through a multistakeholder process. Several other guidelines, such as those of the Carbon Disclosure Project, are compatible with GRI's broader set of criteria. The mission of the GRI is to make sustainability reporting standard practice for all organizations. (Its database includes a growing number of nonprofits and public agencies, including cities. But the data reported here focus as closely as possible on GRI's information on for-profit enterprises.)

The GRI tracks which organizations publicly disclose reports that assess their performance across measures of social, environmental, and economic sustainability, with data back to 1999. In most cases, these are self-assessments that are based on widely accepted measures, in particular the GRI's own guidelines. The number of companies worldwide self-disclosing has grown from 11 in 1999—which included 5 in Europe, 5 in North America, and 1 in Asia—to 2,828 in 2011, the last year for which data have been finalized.[38] The 2011 total included 280 companies in Africa, 641 in Asia, 1,045 in Europe, 326 in Latin America, 414 in Canada and the United States, and 122 in Oceania (including Australia and New Zealand).[39]

It is difficult to estimate the total annual global revenues of businesses that aspire to the new triple bottom line. One indication of the scale currently in play, however, comes from a recent global analysis by the new Global Sustainable Investment Alliance (GSIA). This group has estimated that, excluding Latin America, the investment assets managed with attention to sustainability measures—including screening out for negative environmental, social, and governance factors and screening in on the basis of positive ESG criteria—totaled at least $13.6 trillion by the end of 2011.[40] That represented about 22 percent of all managed investment assets in the regions assessed.[41] For-profit businesses are not the only enterprises that benefit. An estimated $8.3 trillion of the $13.6 trillion is accounted for by negative and exclusionary screening, a strategy to minimize harm.[42] But slightly more than $1 trillion in investment assets was intentionally committed to "positive/best-in-class screening," which the GSIA defined as investments "selected for positive ESG performance relative to industry peers."[43]

Table 2. Cumulative Users of B Lab Assessment Tool and Companies Certified as B Corporations, 2007–13

Year	Users of B Lab Impact Assessment Tool	Companies Certified by B Lab as B Corporations
2007	280	78
2008	855	125
2009	1,744	212
2010	2,976	370
2011	4,834	503
2012	7,240	670
2013*	8,000+	725

*By early April.
Source: B Lab, discussions with author, early April 2013.

Development Aid Falls Short, While Other Financial Flows Show Rising Volatility

Michael Renner and Cameron Scherer

As the world approaches the 2015 deadline for achieving the Millennium Development Goals outlined in 2000, development aid by the 26 members of the Development Assistance Committee (DAC) of the Organisation for Economic Co-operation and Development declined in 2012 for the second year.[1] Preliminary data indicate that official development assistance (ODA) totaled $128.4 billion (in 2011 dollars) that year, down 4 percent from $133.7 billion in 2011.[2] (See Figure 1.) The 2012 figure marks a 6-percent decline from 2010, when global ODA peaked at $136.7 billion.[3]

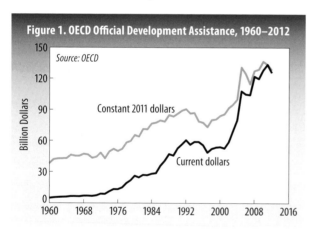

Figure 1. OECD Official Development Assistance, 1960–2012

Source: OECD

Constant 2011 dollars

Current dollars

The United States provided the largest amount of ODA, with a total of $29.9 billion in 2012, which was 23.3 percent of the DAC total.[4] Trailing the United States are the United Kingdom, Germany, France, and Japan.[5] When tracking ODA as a percentage of gross national income (GNI), however, a different picture emerges.[6] Since 1970 the United Nations has set 0.7 percent of GNI as the target for ODA: in 2012, only Luxembourg, Sweden, Norway, and Denmark exceeded this target.[7] (See Table 1.) In comparison, the U.S. figure was only 0.19 percent.[8] Not surprisingly, given the severity of the Eurozone crisis, the 15 European Union members of DAC decreased their assistance by a total of 7.4 percent, with the most severe cuts coming from Spain, Italy, Greece, and Portugal.[9]

It should be noted that DAC governments are not the only ones that provide development assistance: according to a 2012 U.N. report, non-DAC countries donated a total of $7.2 billion in development aid in 2010, with Saudi Arabia providing almost half of the total.[10] Furthermore, assistance from private sources was estimated at $56 billion that same year, but reporting on such flows is much weaker than for government funds.[11]

Humanitarian assistance, or short-term aid provided in response to disasters and humanitarian crises, is a numerically small but highly visible portion of ODA. Preliminary data indicate that, in 2012, assistance provided by governments for such purposes fell 6.5 percent from the previous year, from $13.8 billion to $12.9 billion.[12] (When including non-governmental sources, humanitarian aid fell by 7.7 percent.)[13] This decline is not totally unexpected, as many of the world's leading

Michael Renner is a senior researcher at Worldwatch Institute and codirector of *State of the World 2014.* **Cameron Scherer** is a program associate at Internews.

economies are still recovering from the financial crisis. Also in 2012, the United Nations categorized 76 million people as in need of humanitarian assistance, fewer than the 93 million in 2011.[14]

As with total ODA, the world's top humanitarian donors are not necessarily those that provide the greatest portion of their GNI. The United States once again tops the list of assistance in absolute amounts, with $3.8 billion in 2012, or 29.4 percent of all humanitarian aid (a number that was $483 million below the figure in 2011).[15] Luxembourg (at 0.16 percent of GNI) and Sweden (at 0.14 percent) top the relative standings.[16] In 2012, Turkey played an unusually large role. Its humanitarian assistance surged to $775 million more than in 2011, to $1.04 billion, driven by the country's response to the ongoing conflict in neighboring Syria.[17] In 2011, the most recent year for which data are available, Pakistan, Somalia, and the West Bank and Gaza Strip were the three areas to receive the greatest amount of humanitarian assistance, together taking in over one quarter of global assistance.[18]

But ODA is far from the only mechanism of international capital flows to or from developing countries and emerging markets. Indeed, a multitude of vehicles—private and public, bilateral and multilateral—fill the global finance landscape. And against this broader canvas, ODA involves relatively small amounts of money.

Among public funds, outflows have actually exceeded inflows into developing countries and emerging markets for most of the past decade. (The remaining data in this article are expressed in current dollars as opposed to the ODA data, which are in 2011 dollars.) According to the International Monetary Fund, net outflows of more than $180 billion in 2006 turned into net inflows of close to $140 billion in 2009, but by 2012, there was once again a net outflow of close to $42 billion.[19] (See Figure 2.)

Official flows are dwarfed by private flows, which include direct investments, portfolio investments, and other funds. Whereas official and private net flows were of roughly the same magnitude during the 1980s, private flows began to skyrocket in the past decade, reaching a peak of $691 billion in 2007.[20] But in a sign of unprecedented volatility, net private flows collapsed to just $279 billion in 2008,

Table 1. ODA by Total Amount and Percent of Gross National Income, Selected Countries, 2012*		
Country	ODA Provided	ODA as Share of GNI
	(billion 2011 dollars)	(percent)
United States	$29.9	0.19
United Kingdom	$13.5	0.56
Germany	$14.0	0.38
France	$12.8	0.45
Japan	$10.6	0.17
Sweden	$5.4	0.99
Norway	$4.8	0.93
Denmark	$2.8	0.84
Luxembourg	$0.4	1.00

*In 2011 dollars and exchange rates.
Source: OECD, "Preliminary Data—ODA Data for 2012," April 2013.

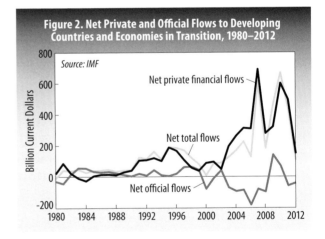

Figure 2. Net Private and Official Flows to Developing Countries and Economies in Transition, 1980–2012

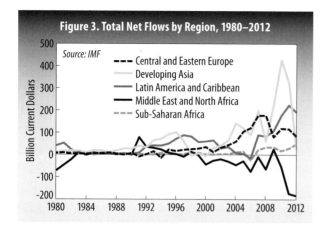

Figure 3. Total Net Flows by Region, 1980–2012

Source: IMF

- - - Central and Eastern Europe
Developing Asia
Latin America and Caribbean
Middle East and North Africa
- - - Sub-Saharan Africa

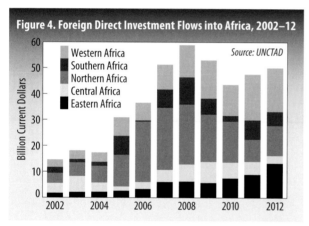

Figure 4. Foreign Direct Investment Flows into Africa, 2002–12

Source: UNCTAD

Western Africa
Southern Africa
Northern Africa
Central Africa
Eastern Africa

surged back to $600 billion in 2010, only to plunge once again to $145 billion in 2012.[21]

Analyzing public and private flows combined, both the order of magnitude and the degree of volatility vary widely among different regions. Net flows into developing Asia—principally China and India—have soared over the past decade, outpacing the flow of funds into Central and Eastern Europe and into Latin America and the Caribbean.[22] But the amounts then plummeted from a peak of $426 billion in 2010 to less than $25 billion in 2012.[23] At the other extreme, the Middle East and North Africa suffered a massive outflow of funds in the past two years, reflecting political instability.[24] Compared with the other regions, sub-Saharan Africa plays a minor role.[25] (See Figure 3.)

Among private flows, the largest amounts are accounted for by foreign direct investment (FDI). Net FDI rose from under $100 billion per year in the 1980s and early 1990s to a peak of $480 billion in 2008.[26] The financial crisis then caused a dip to $335 billion in 2009, but 2011 saw a recovery to $473 billion.[27] The bulk of FDI flowing to developing countries is going to Asia and Latin America.[28] In East Asia and South Asia, almost 90 percent of FDI goes to China and India; in Latin America and the Caribbean, about half goes to Brazil.[29] Only 10 percent of global FDI is destined for Africa.[30]

Yet Africa is currently undergoing tremendous economic growth, being home to 6 of the 10 fastest-growing countries in the past decade. Modest but noteworthy increases in FDI have helped trigger a subsequent increase in regional gross domestic products. In 2002, FDI to the continent stood at just $15 billion; in 2012, it totaled $50 billion, having peaked at $59 billion in 2008.[31] (See Figure 4.) China is behind much of this FDI, providing at least $14.7 billion in 2011.[32]

Private portfolio investments in developing countries and economies in transition involve much smaller amounts than FDI, but these flows show tremendous volatility. From a net outflow of $70 billion in 2008, they went to a peak inflow of $224 billion only two years later.[33] All other forms of private funds have exhibited even larger swings. From a peak inflow of $145 billion in 2007, they turned into record outflows of $466 billion in 2012.[34]

The volatile nature of private financial flows carries tremendous risks for financial and economic stability in recipient countries. This includes concerns about speculative flows, impacts on exchange rates, and the danger of credit, debt, and asset price bubbles. Growing volatility means that there is an increasing risk of sudden and destabilizing withdrawals of capital.[35] The 2012 edition of *World Economic Situation and Prospects* from the United Nations warned that "volatile capital flows originating in the developed economies continue to threaten boom and bust cycles in developing countries."[36]

In general, FDI is more stable than other forms of private investments. Especially where "greenfield" investments in new productive capacity are concerned, FDI is usually undertaken with a longer time horizon in mind, and happens mostly where macroeconomic conditions are stable. In contrast, portfolio investment and cross-border interbank lending are often driven by short-term considerations such as changes in interest rates.[37]

However, the United Nations points to evidence that a growing portion of FDI in recent years is going to investments in financial companies or to intra-company debt. Also, a considerable portion of FDI relates to mergers and acquisitions—and thus represents a transfer of ownership rather than fresh investment.[38] These shifts imply that capital can be moved more easily among countries, and indeed the share of short-term and more-volatile financial FDI flows has increased.[39]

Commodity Supercycle Slows Down in 2012

Mark Konold

Global commodity prices dropped by 6 percent in 2012, a marked change from the dizzying growth during the "commodities supercycle" of 2002–12, when prices surged an average of 9.5 percent a year, or 150 percent over the 10-year period.[1] This change of pace is largely attributed to China's shift to less commodity-intensive growth.[2] Yet while prices declined overall in 2012, some commodity categories—energy, food, and precious metals—continued their decade-long trend of price increases.[3] (See Table 1.)

The commodities market consists of various raw materials and agricultural products with fluctuating value that are bought and sold in global exchanges. This includes agricultural products, such as corn, wheat, soybeans, and cotton; energy sources, such as crude oil and natural gas; metals used in construction, such as copper and aluminum; and precious metals that are often used for financial security, such as gold, silver, and platinum. Commodities categories are not always mutually exclusive: some products (corn, for example) are used as an input for other commodities (such as cattle).

Commodity prices were generally in decline for decades before 2002. But as the number of rapidly growing emerging economies grew after 2000, urbanization led to a surge in demand. Commodities supplies, however, were weak due to under-investment in new capital expenditures as well as the difficulty of procuring new supplies because of factors such as stricter environmental regulations and deposits that were more remote.[4] This opened the door to a dizzying climb in commodities prices over the next 10 years.

During the supercycle, the financial sector took advantage of the changing landscape, and the commodities market went from being little more than a banking service as an input to trading to being a full-fledged asset class—what some people refer to as "the financialization of commodities." These days, large investment banks that participate in both the financial and commercial aspects of commodities trading dominate the landscape.[5] At the turn of the century, total commodity assets under management came to just over $10 billion.[6] By 2008, that number had increased to $160 billion, although $57 billion of that left the market that year during the global financial crisis.[7] The decline was short-lived, however, and by the end of the third quarter of 2012, the total commodity assets under management had reached a staggering $439 billion.[8]

Oil market prices, though still high, were stable in 2012, with the average selling price for crude oil around $105 per barrel, in current dollars.[9] (See Figure 1.) Global supplies of oil grew by 2.5 million barrels per day (mbd), with more than 1 mbd in inventories.[10] Growth was largely attributed to recovered production

Mark Konold is the Caribbean program manager for the Climate and Energy Program at Worldwatch Institute.

Table 1. Annual Commodity Indices Prices, 2002–12						
	Energy	**Food**	**Beverages**	**Fertilizers**	**Raw Materials**	**Precious Metals**	**Metals**
(dollars, 2005 = 100)						
2002	55.43	99.17	99.17	74.43	90.85	80.77	61.39
2003	61.10	100.52	94.53	79.77	96.90	87.29	64.33
2004	74.43	105.13	89.68	87.95	95.68	94.62	82.10
2005	100.00	100.00	100.00	100.00	100.00	100.00	100.00
2006	115.07	108.40	104.57	101.51	115.32	136.82	150.97
2007	120.02	128.19	114.00	137.05	118.49	148.55	171.22
2008	155.78	159.22	129.42	340.63	122.19	168.28	153.95
2009	104.63	142.40	144.03	186.42	117.94	194.31	110.03
2010	128.11	150.18	161.26	165.70	147.21	241.04	159.03
2011	153.60	171.47	169.96	217.88	168.67	303.57	167.70
2012	156.26	176.44	138.61	216.14	137.83	315.40	145.12
(percent)						
Change, 2002–11	177.11	72.91	71.38	192.73	85.66	275.84	173.17
Change, 2011–12	1.73	2.90	−18.45	−0.80	−18.28	3.90	−13.46

Source: World Bank, Annual Commodity Indices Price Data, September 2013, 2005 dollars.

in Libya and increased output in Saudi Arabia and Iraq.[11] Toward the end of 2012, however, global supplies began to contract due to outages in Nigeria, reduced Saudi production, and the effects in Iran of a European Union oil embargo and U.S. economic sanctions.[12] Demand, meanwhile, grew at 1.5 mbd in countries that did not belong to the Organisation for Economic Co-operation and Development (OECD) and declined by 0.6 mbd in OECD countries, a total global growth of only 900,000 barrels per day.[13] Oil use in OECD countries has abated, thanks to higher prices, increased efficiency, and the global recession. And although emerging-market demand has also decreased, it is expected to ramp up to around 0.8 mbd in 2013 as countries such as China and Brazil continue to emerge from the global recession.[14]

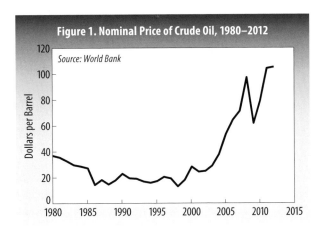

Figure 1. Nominal Price of Crude Oil, 1980–2012

Source: World Bank

The widespread drought in 2012 had an adverse affect on many parts of the agricultural commodities landscape, with corn being hit the hardest. According to the International Monetary Fund, corn is the most vulnerable of crops to price shocks because stocks of the grain remain low.[15] In 2011, corn yields stood at 147 bushels per acre, but in 2012, yields went as low as 122.6 bushels, a 17-percent drop.[16] By year's end, that number had risen only slightly.

According to the U.S. Department of Agriculture, the adverse effects of the drop in corn yields would be felt more strongly in 2013, as supplies would remain tight. The drought in 2012 also had an indirect impact on other important grains, such as wheat. While the wheat harvest was not substantially reduced, the shortage of corn caused consumers to turn to wheat as a substitute. The further tightening of overall supply led to a projected wheat price for 2012–13 of $7.75–$8.45 per bushel, breaking the 2011–12 record of $7.24.[17]

In 2000, countries began reducing grain stocks as years of stable, low food prices and more-liberalized trade eased concerns about food supplies.[18] The resulting low supply levels were swiftly outstripped by growing demand for grains and oilseeds in rapidly growing emerging economies whose populations were suddenly racing up "the food chain," driving prices upward.[19] Thanks to improved supply, agricultural prices have begun to retreat from those levels in recent years.[20] But the cost of inputs remains high and so, therefore, do prices.[21]

Further complicating price dynamics in the agricultural sector is a crop's end use, especially whether it is used for food or biofuels. According to a 2011 report by the Farm Foundation, global corn use in the category "food, seed, and industrial" has expanded by 88 percent since the 2005–06 marketing year.[22] Ethanol falls in this category. During this time in the United States, use of corn increased by 2.23 billion bushels, and corn usage for ethanol increased by 2.46 billion bushels.[23] The strong demand put upward pressure on corn prices and on the price of other commodities displaced by the expanded area devoted to corn production.

Throughout the commodities supercycle, the price of precious metals grew robustly. Although the rate of price increases for some precious metals slowed recently, gold maintained its momentum in 2012. In constant 2012 dollars, the average annual price for gold was $1,669 per troy ounce (oz t.), a 3.9-percent increase from 2011.[24] (See Figure 2.) A large factor in gold's increase in value was purchases by central banks. For the eighth year in a row, central banks were net buyers of gold, with purchases in 2012 up 17 percent from 2011.[25] Other precious metals saw substantial declines in price. Silver fell 13 percent to $31.15 per oz t., platinum dropped 11.7 percent to $1,551 per oz t., and palladium fell 14 percent to $643 per oz t.[26]

Prices of other metals and minerals sustained high levels in 2012, largely because of continued high levels of consumption. Despite price reductions for aluminum in large economies like the European Union and Brazil of 7.7 and 5.2 percent, respectively, the metal's price surged in economies like India and China, 7.5 and 15 percent, respectively, for an overall global increase of 6.8 percent in one year.[27] High prices were also sustained because of supply constraints. Aluminum supplies grew by only 3.2 percent and copper grew by just 4.4 percent.[28] One of the few metals to see significant growth in availability was nickel, which saw a 13 percent boost in supplies.[29]

Stockpiling by owners was another reason for sustained metal prices in 2012.

As the demand for metals declined at the start of the financial crisis in 2008, owners set aside inventory to prop up prices. In one year, metal inventories registered on the London Metal Exchange jumped by 313 percent.[30] As a result, banks and trading companies began buying up warehouse space and storing surplus supplies, thereby inflating the prices of metals.[31]

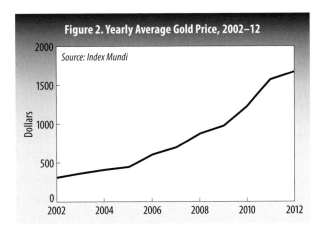

Figure 2. Yearly Average Gold Price, 2002–12

Source: Index Mundi

The slowdown in commodity price growth in 2012 was indeed notable, but it is still not clear if the supercycle is completely over. Prices are still much higher than they were in 2002, but the dramatic slowdown in Chinese demand has investors abandoning these markets. By the end of April 2013, the commodities markets saw a loss of $63 billion, and Barclays Bank claimed that total commodity assets under management had dropped to their lowest level in three years.[32] It is going to take a little more time to find out whether the commodities market has permanently cooled, reverses dramatically, or picks up and resumes its blistering pace.

Peace and Conflict Trends

Burundi peacekeepers prepare for their next rotation to Somalia

For additional peace and conflict trends, go to vitalsigns.worldwatch.org.

Military Expenditures Remain Near Peak

Michael Renner

In 2012, world military expenditures ran to $1,740 billion, expressed in constant 2011 dollars ($1,753 billion in current prices).[1] According to the World Military Expenditure Database of the Stockholm International Peace Research Institute (SIPRI), this is just slightly below the peak value of $1,749 billion in 2011, but still higher than in any other year since the end of World War II.[2]

A lack of transparency limits data availability for some countries. Excluding the four countries for which there are no consistent data over the years (Afghanistan, Honduras, Iraq, and Qatar), SIPRI offers a time series of global military spending for the past 25 years. After the end of the cold war, spending declined by about one-third, from $1,613 billion in 1988 to $1,053 billion in 1996.[3] But it did not take long for budgets to bottom out and grow again. Starting in 1998 and particularly following the September 2001 attacks in the United States, military budgets were resurgent, expanding by 65 percent.[4] (See Figure 1.)

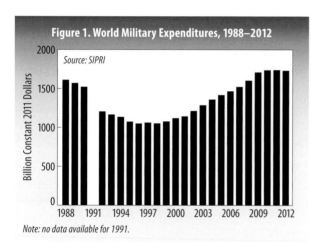

Figure 1. World Military Expenditures, 1988–2012

Source: SIPRI

Note: no data available for 1991.

A variety of factors drive military expenditures, although the precise circumstances and motivations differ substantially across the world. In some countries, warfare—against either a neighboring country or a domestic opponent—is the key driver. Other countries maintain considerable military establishments even though they face little prospect of attack. Deterrence is often invoked as a key reason, but some countries—like the United States, the United Kingdom, and France—routinely intervene in the affairs of other countries.

The manufacturers of armaments—producing anything from "conventional" weapons like tanks, missiles, and fighter jets to nuclear weapons and drones, as well as related services—are powerful proponents of large military spending. In current dollar terms, sales by the world's leading 100 armaments companies more than doubled in the last decade to $410 billion in 2011, but after inflation they rose by 51 percent.[5] (See Figure 2.) In 2011, U.S. companies accounted for close to 60 percent of these sales; West European firms had 29 percent of sales and Russian firms had 3.5 percent, with companies from all other countries (except China, for which comparable and accurate information is not available) contributing just 7.8 percent ($31.3 billion).[6]

Michael Renner is a senior researcher at Worldwatch Institute and codirector of *State of the World 2014.*

During 2012, military budget trends varied widely among different countries and regions. In absolute terms, Russia (spending $12.3 billion more than in 2011), China ($11.4 billion), and Saudi Arabia ($5.7 billion) revved up their budgets the most.[7] Also prominent among the countries that increased their expenditures were Oman ($2.2 billion), Indonesia ($1.3 billion), and Colombia ($1.1 billion).[8]

By contrast, the biggest decrease in expenditures occurred in the United States ($42.6 billion less).[9] A large number of European countries cut their budgets due to economic distress, including Italy ($2 billion), Spain ($1.8 billion), the Netherlands ($900 million), and Portugal ($800 million).[10] Among the other significant cutters were India ($1.4 billion)—which reversed its previously rapid pace of increased spending—Australia ($1.1 billion), and Canada ($900 million).[11]

But in relative terms, other countries stand out. Zimbabwe raised its military spending by 53 percent, followed by Oman (51 percent), Paraguay (42 percent), Venezuela (39 percent), and Kazakhstan (30 percent).[12] The biggest cutters were Uganda (–57 percent), South Sudan (–42 percent), and Hungary (–20 percent).[13]

For the past quarter-century, the United States has been the world's military colossus. During those years, it routinely accounted for more than 40 percent of global expenditures, but in 2012, that figure declined to 39 percent.[14] Nonetheless, no other country even comes close. At $682.5 billion in 2012 (current dollars), the United States spends as much as the next 11 countries combined ($683.4 billion), with the rest of the world beyond these 12 accounting for a comparatively small $387.3 billion.[15] (See Figure 3.)

Following the United States, the largest spenders are China at an estimated $166 billion, Russia ($91 billion), the United Kingdom ($61 billion), Japan ($59 billion), France ($59 billion), Saudi Arabia ($57 billion), India ($46 billion), Germany ($46 billion), Italy ($34 billion), Brazil ($33 billion), and South Korea ($32 billion).[16] So large a spender is the United States that the reduction in its 2012 budget was almost equal to the entire military budget of the ninth-largest spender, Germany.[17]

Since the end of World War II, U.S. military spending has gone through tremendous up and down swings, with peaks marked by wars in Korea and Vietnam in the early 1950s and late 1960s, followed by a buildup near the end of the Cold

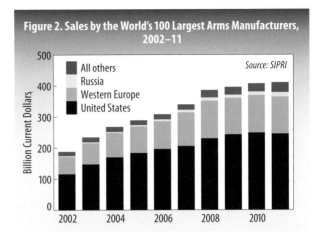

Figure 2. Sales by the World's 100 Largest Arms Manufacturers, 2002–11

Source: SIPRI

All others
Russia
Western Europe
United States

Billion Current Dollars

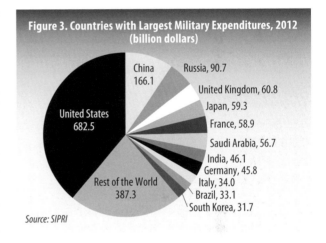

Figure 3. Countries with Largest Military Expenditures, 2012 (billion dollars)

China 166.1
Russia, 90.7
United Kingdom, 60.8
Japan, 59.3
France, 58.9
Saudi Arabia, 56.7
India, 46.1
Germany, 45.8
Italy, 34.0
Brazil, 33.1
South Korea, 31.7
United States 682.5
Rest of the World 387.3

Source: SIPRI

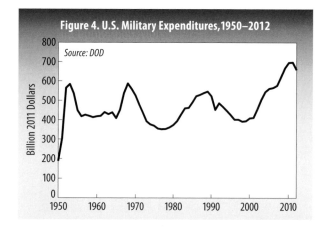

Figure 4. U.S. Military Expenditures, 1950–2012

Source: DOD

(Y-axis: Billion 2011 Dollars; X-axis: 1950–2010)

Table 1. U.S. Military Deployments Worldwide, July 2013	
Region	**Military Personnel**
United States and Territories	1,208,083
Europe	69,065
East Asia and Pacific	51,259
Undistributed *	37,352
Middle East & North Africa	4,389
Central and South America	1,788
Sub-Saharan Africa	269
Former Soviet Union	78
South Asia	51
Total	1,372,334

*Afghanistan, Iraq, Kuwait, South Korea, and undisclosed locations.
Source: U.S. Department of Defense, DMDC, "Active Duty Military Personnel by Service by Region/Country," as of 31 July 2013.*

War in the late 1980s, and most recently the massive expansion undertaken in the name of the war on terror and the invasions of Afghanistan and Iraq.[18] (See Figure 4.)

Unparalleled in the world, the United States maintains bases or has some other military presence in almost every country. As of July 2013, the country had about 164,000 soldiers stationed abroad, in addition to 1.2 million military personnel at home and in U.S. territories.[19] (See Table 1.) Close to 140,000 people were involved in "contingency operation deployments" related to the wars in Afghanistan and Iraq as of mid-2013.[20]

Countries in conflict-ridden areas are among those that spend the most on the military relative to their economic means. Saudi Arabia tops the list with 8.9 percent of gross domestic product (GDP) absorbed by military budgets in 2012, with South Sudan and Oman (each at 8.4 percent) and Israel (6.2 percent) close behind.[21] The United States and Russia both spend 4.4 percent of their GDP on the military, whereas India (2.5 percent) spends at the global average level.[22] SIPRI estimates China's share at 2 percent.[23]

At a time when endemic poverty, mass unemployment, health epidemics, and the looming threats of climate change cry out for greater attention, the continued largesse for military purposes in many countries reflects a troubling set of priorities.[24]

In a world where 2.4 billion people struggle to survive on $2 per day or less (hardly changed from 2.6 billion in 1981), in 2012 governments spent on average $249 on weapons and soldiers for each person on the planet.[25] The $1,234 billion that high-income countries spent on military programs in 2012 is more than nine times the $128.4 billion they allocated for development assistance.[26]

As rich-poor divides widen in wealthy countries, military priorities there deserve greater scrutiny. According to the Friends Committee on National Legislation, 37¢ out of each dollar of U.S. federal government spending in 2012 went to pay for past and current wars.[27] By contrast, 19¢ went to health care, 15¢ to anti-poverty efforts, 6¢ to jobs and economic development, 3¢ to energy, science, and environment, 2¢ to education and social programs, and another 2¢ to diplomacy, international aid, and support for the United Nations and other international agencies.[28]

Although it may now have peaked, the surge in military budgets is part of a larger worrisome trend toward increasing spending on a wide range of measures undertaken in the name of security and on closely entwined issues such as counter-terrorism and anti-narcotics campaigns. This includes expanding internal security bureaucracies (such as the Department of Homeland Security in the United States), the militarization of civilian police departments, and ever-expanding intelligence and surveillance efforts directed at political adversaries, economic competitors, and civilian populations alike.

For most countries, and especially for Russia and China, there is no publicly available information about intelligence budgets and similar measures. But the United States likely devotes the most resources of any country to such purposes. The country's 16 intelligence agencies requested a combined budget of $52.6 billion for 2013, down from the 2011 peak of $54.6 billion.[29] The agencies employ more than 107,000 people (including about 84,000 civilian government employees and more than 23,000 military personnel) and almost 22,000 private contractors.[30] A *Washington Post* investigation found that some 854,000 people have top-secret government security clearances in the United States and that analysts publish as many as 50,000 intelligence reports each year.[31]

Governments have created a large and well-funded apparatus of security agencies, but in numerous ways have failed to address many of the underlying reasons for the world's conflicts and instabilities.

Peacekeeping Budgets Equal Less Than Two Days of Military Spending

Michael Renner

The approved budget for United Nations peacekeeping operations from July 2013 to June 2014 runs to $7.83 billion—$390 million higher than in the previous year.[1] (See Figure 1.) This is the third-highest budget since the record $8.26 billion spent in 2009–10.[2] Despite some relatively minor fluctuations in the last seven years, peacekeeping budgets are much more stable now than in the 1990s, when a rapid rise in spending was followed by an abrupt decline.

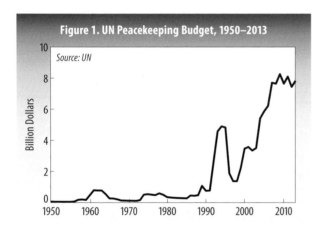

Figure 1. UN Peacekeeping Budget, 1950–2013

Source: UN

Since peacekeeping was invented in the years following World War II, the United Nations has spent a cumulative $124 billion on these missions—an amount that pales in comparison to even a single year of world military expenditures, which stood at $1,753 billion in 2012.[3] The world's armies could not operate for even two days on the current annual peacekeeping budget.

Compared with the early days of peacekeeping—when missions were largely limited to monitoring and maintaining peace along well-defined ceasefire lines—today's missions are highly complex. Some attempt peace enforcement (suppressing the use of violence by combatants), and some are charged with disarming and reintegrating former fighters. Others involve a broad array of civilian tasks, such as assistance in elections and other political processes, institution building, reform of judicial systems, and training for police forces, as well as other steps to foster and consolidate peace.

As of January 2014, the United Nations maintained 15 peacekeeping missions: 8 in Africa, 3 in the Middle East, 2 in Asia, and 1 each in Europe and the Americas.[4] The organization deployed a total of 117,630 personnel, most of whom are uniformed peacekeepers.[5] (See Table 1.) For comparison, three dozen countries have armies that surpass the strength of uniformed and civilian peacekeepers, and five countries have more than 1 million soldiers each.[6]

Following a buildup of uniformed peacekeepers during the early 1990s to about 78,000 (in 1994), the numbers dropped precipitously to fewer than 19,000 in 1999, as some missions came to an end and as enthusiasm for peacekeeping cooled in the wake of difficulties and failed missions.[7] But the numbers soon began to climb again, reaching new heights. In March 2010, the number of troops,

Michael Renner is a senior researcher at Worldwatch Institute and codirector of *State of the World 2014*.

military observers, and civilian police reached a new peak of 101,939.[8] The number has remained at just below the 100,000 mark since then.[9] (See Figure 2.)

The number of civilians involved in peacekeeping is smaller, accounting for between 15 and 19 percent of all peacekeepers in the last decade.[10] The number grew from 8,430 in January 2000 to a peak of 22,616 by December 2010, but then fell below 19,000 in the last two years.[11]

In addition to peacekeeping operations, there are also several "political and peacebuilding" missions with mostly civilian staff. In January 2014, there were 13 such missions (typically, follow-up efforts once a peacekeeping mission comes to a close) in Africa, the Middle East, and western Asia, with a total of 3,810 personnel.[12] With the exception of the U.N. Assistance Mission in Afghanistan and the Assistance Mission for Iraq, these are very small deployments.[13]

During six-and-a-half decades of peacekeeping, 3,211 peacekeepers have been killed on duty, including more than 1,400 fatalities in the currently active missions.[14] Almost 40 percent of all fatalities occurred just during the last decade, the most dangerous one for peacekeepers so far (although the largest number of deaths in a single year occurred in 1993).[15]

A number of peacekeeping missions have been active for decades, providing stability even if not a fundamental resolution to the underlying conflicts. Two of the currently active missions (in the Middle East and South Asia) were established in the late 1940s, one (Cyprus) in the 1960s, two (Syria and Lebanon) in the 1970s, and two (Western Sahara and Kosovo) in the 1990s.[16] The remaining eight operations have been initiated since 2003.[17] Four of these began in the last four years—in the Democratic Republic of Congo (DRC) (2010), in Abyei (a part of Sudan bordering on South Sudan) and in South Sudan (both in 2011), and in Mali (2013).[18]

In January 2014, the largest missions, in terms of both expenditures and personnel, were operations in the DRC (with 25,723 people and a budget of $1.5 billion) and in Darfur, Sudan (with 23,754 people and a budget of $1.3 billion).[19] The South Sudan mission has the next largest budget, at more than $900 million, and more than 10,000 personnel.[20] Operations in Lebanon, Côte d'Ivoire, Liberia, Mali, and Haiti deploy between 9,000 and 11,400 people each and have budgets ranging from $476 million to $602 million.[21] The figures for the remaining seven missions are much smaller.

Table 1. U.N. Peacekeeping Personnel, by Category, January 2014	
Soldiers	83,702
Police	13,180
Military Observers	1,857
Sub-Total, Uniformed	98,739
Civilians, international	5,190
Civilians, local	11,698
U.N. Volunteers	2,003
Sub-Total, Civilian	18,891
Total	117,630

Sources: UNDPKO, "Troop and Police Contributors"; UNDPI, "United Nations Peacekeeping Operations. Fact Sheet: 31 January 2014."

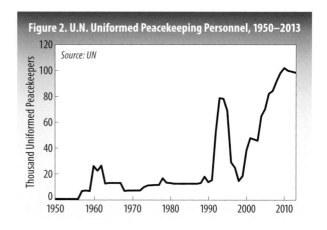

Figure 2. U.N. Uniformed Peacekeeping Personnel, 1950–2013

Source: UN

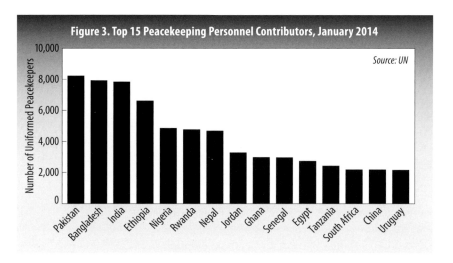

Figure 3. Top 15 Peacekeeping Personnel Contributors, January 2014

Peacekeepers come from all corners of the world. Altogether, 123 countries contributed personnel to these U.N. missions in January 2014—including many nations that themselves have suffered from armed conflict.[22] The top 15 contributors (each of which made at least 2,000 peacekeepers available) provided 65,887 peacekeepers, or about half of the total.[23] (See Figure 3.) With very few exceptions, these are all poor countries. South Asia is particularly prominent: Pakistan, Bangladesh, India, Nepal, and Sri Lanka together accounted for 29,764 peacekeepers, or 30 percent of the total.[24]

Decisions to establish, expand, or terminate a peacekeeping operation are taken by the members of the Security Council. But among the five permanent members, only China is making a substantial personnel contribution with 2,186 peacekeepers. France is second in this group (957), while the United Kingdom (282), the United States (114), and Russia (103) are barely visible.[25] Together, the "Perm-5" account for less than 4 percent of total U.N. peacekeeping forces.[26]

To finance the missions, all U.N. member states are assessed a portion of the total costs according to a formula that includes ability to pay, along with other criteria. The top 10 contributors (which include the Perm-5) provided 80.4 percent of the total budget in 2013, with the United States and Japan alone accounting for almost half of that (39 percent).[27] The next four—France, Germany, the United Kingdom, and China—contributed nearly 28 percent, while Italy, Russia, Canada, and Spain paid another 13.5 percent.[28]

Member states do not always pay in full or on time, leading to repeated arrears. Like the peacekeeping budgets, arrears remained small until the early 1990s. They crossed the $1 billion line in 1992 and have not fallen below that since then.[29] Heavily fluctuating from year to year, they reached a peak of close to $3.5 billion at the end of 2007 and stood at $2.2 billion at the end of 2013.[30]

Nowadays, the United Nations is far from the only organization that dispatches peacekeepers. Various types of non-U.N. missions can be found in all regions of the world, although they sometimes work in conjunction with the Blue Helmets of the United Nations.

During 2013, some 17 missions directed by the European Union (EU) were active, including 9 in Africa, 3 in the Middle East, 4 in Europe, and 1 in Asia (Afghanistan).[31] The goals included peacekeeping with the help of soldiers, securing borders and other monitoring functions, policing, and ensuring the rule of law.[32] The Organization for Security and Co-operation in Europe (OSCE) maintained 16 operations in Eastern Europe, the Caucasus, and Central Asia.[33] NATO runs KFOR (Kosovo Force) in Kosovo and the International Security Assistance Force (ISAF) in Afghanistan, but ISAF in particular is more aptly described as a combat operation than a peacekeeping mission.[34] Four operations are managed by the African Union and African regional organizations, with a Somalia mission by far the largest one.[35] Finally, there were eight missions by various other organizations or coalitions of the willing, in Moldova, the Philippines, the Solomon Islands, the Sinai and the West Bank, Côte d'Ivoire, Mali, and Colombia.[36]

Altogether, U.N. and non-U.N. missions deployed about 251,000 people in 2013, about 213,000 of whom were military personnel.[37] (See Table 2.) Military contingents account for 85 percent of the total personnel; excluding the two NATO deployments lowers the share to 75 percent.[38] The African and "other" missions also are heavily military-dominated, but military personnel make up a much smaller 58 percent of EU missions, and the OSCE missions do not have any military functions at all.[39] Except for the EU deployments and some U.N. peacekeeping missions, police contingents play a very minor role.

Peacekeeping missions of various stripes have become a steady presence in

Table 2. U.N. and Non-U.N. Peacekeeping Missions, 2013

Organization	Number of Missions	Personnel			
		Military	Police	Civilian	Total
U.N. Peacekeeping	15	84,851	13,053	18,851	116,755
U.N. Political and Peacebuilding	13	360*	–	3,450	3,810
European Union	17	2,891	932	1,191	5,014
OSCE	16	0	13	457	470
NATO	2	91,138	56	998	92,192
African organizations	4	23,750	1,629	193	25,572
Others	8	6,860	186	147	7,193
Total	75	213,091	15,869	25,287	251,006
Total, excluding NATO	73	118,712	15,813	24,289	158,814

*All uniformed personnel.
Note: Data for U.N. missions are for year-end 2013. Data for all other missions are for mid-year 2013.
Sources: UNDPI, "United Nations Peacekeeping Operations. Fact Sheet: 31 December 2013" and "Fact Sheet: 31 January 2014"; ZIF, "International and German Personnel in Peace Operations 2013–14," September 2013.

several regions of the world. While they are not necessarily making the world a more peaceful place—with the exception of the small OSCE missions, they are only dispatched after a conflict weighs on the conscience of global decision makers—they fulfill an important role in trying to bring a semblance of peace and order to troubled areas.

Population and Society Trends

Mobiles and more; Stung Treng, Cambodia

For additional population and society trends, go to vitalsigns.worldwatch.org.

Displaced Populations

Michael Renner

For reasons that range from warfare and persecution to natural disasters and development projects, an estimated 92.5 million people were forcibly displaced in 2012, either inside their home countries or across a border.[1] (See Figure 1.) Displacement is sometimes temporary, but in other cases it can last for years.

International refugees under the care of the U.N. High Commissioner for Refugees (UNHCR) numbered 10.5 million, and there are also close to 1 million asylum seekers worldwide.[2] Meanwhile, the U.N. Relief and Works Agency for Palestine Refugees in the Near East (UNRWA) is tasked with providing support to about 5 million Palestinian refugees.[3] Internally displaced persons (IDPs)—at 28.8 million—outnumber international refugees by a significant margin.[4] People displaced by natural hazards—typically also displaced inside their own countries but seen as a separate category since, unlike IDPs, they are not victims of human actions—ran to more than 32 million in 2012, but this number varies considerably from year to year.[5] In addition, a large number of people are displaced by ill-considered development projects. No firm numbers exist, but the *World Disasters Report 2012* offers a rough guess of 15 million such individuals.[6]

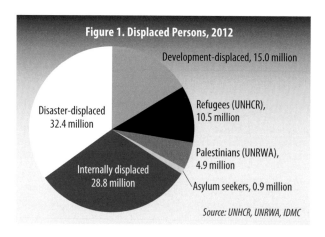

Figure 1. Displaced Persons, 2012

Development-displaced, 15.0 million

Disaster-displaced 32.4 million

Refugees (UNHCR), 10.5 million

Palestinians (UNRWA), 4.9 million

Internally displaced 28.8 million

Asylum seekers, 0.9 million

Source: UNHCR, UNRWA, IDMC

The number of people who leave involuntarily—the 92.5 million people just described—remains considerably lower than that of people who leave of their own volition.[7] Long-term international migrants—people who decide to live outside their home country for a year or longer—are estimated at 214 million, and internal migrants may number as many as 740 million.[8] The ranks of both groups of migrants have grown significantly over the past half-century as economies have become more interconnected.[9]

Over time, the trends in refugee and IDP numbers have diverged substantially.[10] (See Figure 2.) The number of international refugees climbed from below 5 million in the 1950s, 1960s, and 1970s to a peak of close to 18 million in the early 1990s, but it has since declined to just above 10 million.[11] Given the unresolved Israeli-Palestinian conflict, the number of Palestinian refugees has been rising over the decades from less than 1 million in 1950 to about 5 million today.[12] For internally displaced persons, only a much shorter time series of data is available. Following a rapid rise and then decline in the 1990s, their numbers have risen steadily to record levels.[13]

Michael Renner is a senior researcher at Worldwatch Institute and codirector of *State of the World 2014*.

More than half of all refugees worldwide come from just five countries. Afghanistan (at 2.6 million) remains the leading country of origin for refugees, followed by Somalia (1.1 million), Iraq (746,000), Syria (729,000), and Sudan (569,000).[14] Developing countries also host the vast majority of international refugees. By far the largest number of these displaced people have found refuge in Pakistan (1.6 million), followed by Iran (868,000), Germany (590,000), and Kenya (565,000).[15] Relative to the total host population, however, Jordan has the largest number of refugees on its territory (49 per 1,000 inhabitants), followed by Chad (33) and Lebanon (32).[16]

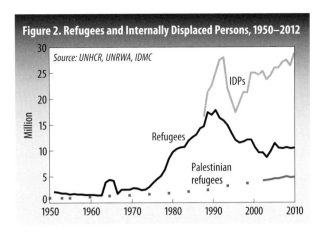

Figure 2. Refugees and Internally Displaced Persons, 1950–2012

During 2012, some 1.1 million people were newly displaced across international borders—principally from the Democratic Republic of the Congo (DRC), Mali, Somalia, Sudan, and Syria.[17] This is equivalent to about 3,000 persons each day, and it is the highest annual figure since 1999.[18] Meanwhile, only 526,000 refugees returned to their home countries voluntarily during 2012—primarily to Afghanistan, Côte d'Ivoire, and Iraq.[19]

Large numbers of people are fleeing conflict zones because the distinction between combatants and civilians is often blurred in many contemporary conflicts. Armed groups may not feel any responsibility toward local populations, and certain groups of civilians are even deliberately targeted.

Around 6.5 million people were newly displaced inside their own countries during 2012, far surpassing returnees and thus raising the total number of IDPs to the highest figure since comparable data were recorded in 1989.[20] Colombia has the largest number of IDPs, estimated at between 4.9 million and 5.5 million, followed by Syria (3 million), the DRC (2.7 million), Sudan (2.2 million), and Iraq (2.1 million).[21] Regionally, sub-Saharan Africa has the largest number of IDPs at 10.4 million, followed by the Middle East and North Africa (6 million), the Americas (5.8 million), South and Southeast Asia (4.1 million), and Europe/Central Asia (2.5 million).[22]

Unlike international refugees, many IDPs do not receive protection and assistance, often because their own governments or armed opposition groups block access to them. Lack of safety for aid workers and inadequate transport and logistical infrastructure are additional factors. UNHCR has had an expanded role in caring for IDPs, and the number that agency has under its care has risen from 4–5 million people a decade ago to some 17.7 million people at the end of 2012—about 60 percent of the worldwide IDP population.[23]

UNHCR's responsibility has steadily expanded over the years, reflecting greater complexity in the types and patterns of displacement around the world. This concerns not only IDPs but also stateless people as well as people affected by major natural disasters.[24] The number of people forced to flee in the face of rapid-onset natural disasters fluctuates strongly from year to year, depending, of course, on the

Table 1. People Displaced by Natural Disasters, 2007–12

Year	Displaced Persons
	(million)
2007	25.0
2008	36.1
2009	16.7
2010	42.3
2011	16.4
2012	32.4

Sources: Data for 2007 from UNHCR, 2007 Global Trends: Refugees, Asylum-seekers, Returnees, Internally-Displaced and Stateless Persons (Geneva: 2008), p. 2; other data from IDMC, Global Estimates 2012. People Displaced by Disasters (Geneva: 2013), p. 11.

severity and frequency of disasters. (See Table 1.) The figures include people displaced by floods, storms, earthquakes, and wildfires but not droughts, which also may generate large movements of people. In some cases, people are displaced for prolonged periods of time, especially when there are recurring disasters.[25]

In 2012, three quarters of all disaster-related displacements took place in just four countries: more than 9 million people in India, 6.1 million in Nigeria, more than 5.5 million in China, and 3.7 million in the Philippines.[26] Hurricane Sandy, by comparison, displaced 1.1 million people in the United States and Cuba.[27]

In the five-year period of 2008–12, China suffered the largest number of displacements (49.8 million), followed by India (23.8 million), Pakistan (15 million), and the Philippines (12.3 million).[28] Developing countries accounted for a staggering 98 percent of all displacements in this period.[29] Relative to total population, Haiti, Chile, the Philippines, and Pakistan had the largest displacements—more than 1 percent of their populations in each case.[30] During those five years, some 29 mega-events (defined as displacing at least 1 million people each) accounted for 68 percent of all displacements; large events (between 100,000 and 1 million people displaced) accounted for another 26 percent.[31] Medium to small events have just 6 percent of the total, but it is likely that this figure is underreported.[32]

The Internal Displacement Monitoring Centre (IDMC) in Geneva expects that disaster-related displacement will rise in coming years due to a number of factors that include population growth in hazard-prone areas, the growth of substandard housing, and increasingly frequent and intense weather events due to climate change.[33]

Indeed, the number of climate-displaced persons is generally expected to rise in coming years as extreme weather events become more frequent and intense and as droughts, desertification, sea level rise, and glacier melt become more prominent phenomena. The International Organization for Migration, for example, has suggested that in a world that is 4 degrees Celsius warmer, the commonly cited estimate of 200 million people displaced by climate change by 2050 could "easily be exceeded."[34] But it is impossible to make any reliable projections about how many people may be uprooted due to climate change in coming years and decades. There are too many unknowns to be able to predict the scale of population movements to come, let alone their direction, destination, and timing.[35]

According to UNHCR, "global social and economic trends indicate that displacement will continue to grow in the next decade, exacerbated by population growth, urbanization, natural disasters, climate change, rising food prices and conflict over scarce resources."[36] Yet "people who are displaced across borders owing to natural disasters and the effect of climate change face a potential legal protection gap, since they are not covered by the 1951 UN Refugee Convention."[37]

For development-induced displacement, the true numbers remain largely unknown, especially given year-to-year fluctuations. The estimate of 15 million mentioned in the *World Disasters Report* may well be a substantial understatement.[38] In a 2007 report, Christian Aid offered a much higher estimate of 105 million people.[39]

"Development" has positive connotations, but people affected by it suffer the adverse consequences of projects like road building and the construction of dams, industrial facilities, or biofuels plantations on land they inhabit. Project planning and decision making often take place without the input or even the consent of affected communities. As the International Federation of Red Cross and Red Crescent Societies points out, people uprooted by development projects—unlike those displaced by conflicts—almost always remain within their own countries.[40] But they typically suffer permanent rather than temporary displacement.

There is growing recognition that it will be increasingly difficult to easily categorize people displaced by separate causes. Environmental problems are often closely intertwined with socioeconomic conditions such as poverty and inequality of land ownership, resource disputes, poorly designed development projects, and weak governance. Distinguishing between forced and voluntary movements of people in a clear-cut way is becoming harder. Instead of distinctions written in stone, it is more useful to think in terms of a continuum of causes and factors. Indeed, as the 2012 edition of the *World Disasters Report* explains, the term "mixed migration" is increasingly being used.[41] For a better understanding of the dynamics and for more productive discussions about possible policies, it is important for migration, refugee, and environmental experts to work together to address the full continuum.

World Population: Fertility Surprise Implies More Populous Future

Robert Engelman

World population reached 7.2 billion in mid-2013, according to United Nations demographers, with present and projected future growth propelled in part by unexpectedly high fertility in a number of developing countries.[1] Based on current trends in global birth, death, and migration rates, the United Nations projects a variety of future population scenarios, with the three principal ones suggesting that world population will be somewhere between 6.8 billion and 16.6 billion at the end of this century.[2] (See Figure 1.) Using a number based literally on a projection of trends through 2010, the U.N. demographers determined that 82.1 million people were added to the world's population in 2012—the highest annual increment since 1994.[3] (See Figure 2.)

Based in large part on the 2010 round of annual censuses in countries around the world, the new U.N. Population Division report, *World Population Prospects: The 2012 Revision*, dispels a widespread view that experts expect population growth to end "on its own" sometime in the second half of the twenty-first century.[4] Rather, the new medium-fertility or best-guess scenario suggests the most likely outcome is that world population will continue to grow throughout this century and into the next. In this scenario, the world still gains more than 10 million people in the year 2100 and closes the century at 10.9 billion.[5]

By 2050, the year when many in the environmental and food security fields had been assuming the world will be home to around 9 billion human beings, the new projections suggest instead a global population of 9.6 billion.[6] That is about 700 million people more than the 8.9 billion the U.N. Population Division had projected for 2050 just 10 years ago.[7]

The report, part of a series updated every two years, details the U.N. Population Division's estimates of population size, growth, fertility, age structure, and related dynamics from 1950 to 2010 in all countries and regions.[8] Separately, the report details projections from 2010 through 2100. Unlike estimates, projections are conditional forecasts based on current data and assumptions about how the key demographic forces of birth, death, and migration might evolve.

While the biggest surprise in the report came from the projections of faster future population growth than had been expected, these numbers actually have their roots in a surprise about the present: Women in many developing countries are having more children today than U.N. demographers previously thought.[9] Indeed, the authors reported that they had raised by a full 5 percent their estimates of current fertility in 15 sub-Saharan African countries—including Nigeria, Niger, Ethiopia, and the Congo—where family size is already among the highest in the world.[10]

Robert Engelman is a senior fellow at Worldwatch Institute. Janice Pratt provided research and data assistance on this article.

In its 2010 series of population estimates and projections, for example, the Population Division projected Burundi's fertility rate to be 4.1 children per woman in the five years from 2010 to 2015; the Division now projects that number to be 6.1 children.[11] (Even the newer series considers data for this period to be projections, since the latest actual data come from 2010, with limited or no data in most countries from 2011 through 2013.)[12] For Ethiopia, the fertility projection for this period rose from 3.9 to 4.6 children per woman, while for Mali it rose from 6.1 to 6.9.[13] Population Division director John Wilmoth said that some of these higher estimates resulted from recent increases in fertility that had not been detected until now, while others were based purely on reassessments of past and existing data.[14]

While the reasons behind the higher-than-expected fertility in many countries are not fully understood, they correlate well with recent government reluctance to give priority to and fund family planning services in some of the world's poorest countries.[15] Spending on family planning services in developing countries both by those countries' governments and by wealthier donor governments and intergovernmental agencies has stagnated in recent years at around $4 billion annually.[16] More than twice that is needed to reach the estimated 222 million women who are sexually active and do not want to become pregnant but are not using contraception.[17] Research suggests that stagnant spending on family planning may be contributing to higher levels of unintended pregnancy.[18] About two out of five pregnancies worldwide are unintended—in industrial countries as well as developing ones—and more than one in five births worldwide results from such pregnancies.[19]

The new report estimates that the world's population is growing at about 1.15 percent annually, and—despite the higher-than-anticipated fertility in many countries—that the growth rate is continuing to slow.[20] Most human beings—an even 60 percent—live in Asia, with Africa the second most populous region, followed by Europe, Latin America and the Caribbean, North America, and Oceania.[21] (See Figure 3.)

Population growth, however, is changing this distribution: Approximately 96 percent of the growth is occurring in developing countries, with Asia accounting for 54 percent of growth.[22] (See Figure 4.) Africa, growing more rapidly than Asia but from a smaller base of population, accounts for almost one-third of current

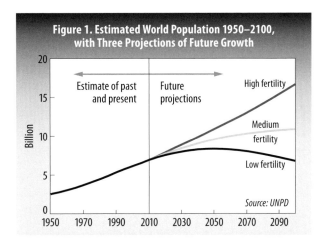

Figure 1. Estimated World Population 1950–2100, with Three Projections of Future Growth

Source: UNPD

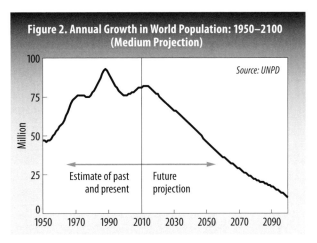

Figure 2. Annual Growth in World Population: 1950–2100 (Medium Projection)

Source: UNPD

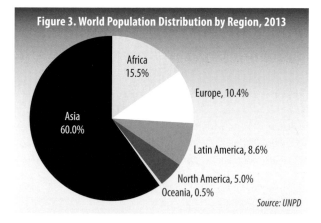

Figure 3. World Population Distribution by Region, 2013

Africa 15.5%
Europe, 10.4%
Asia 60.0%
Latin America, 8.6%
North America, 5.0%
Oceania, 0.5%
Source: UNPD

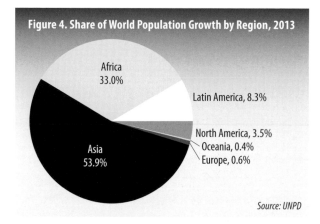

Figure 4. Share of World Population Growth by Region, 2013

Africa 33.0%
Latin America, 8.3%
North America, 3.5%
Oceania, 0.4%
Europe, 0.6%
Asia 53.9%
Source: UNPD

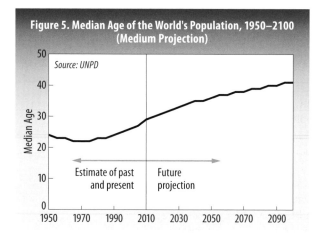

Figure 5. Median Age of the World's Population, 1950–2100 (Medium Projection)

Source: UNPD
Estimate of past and present
Future projection
Median Age

growth.[23] Europe, many of whose countries are experiencing shrinking populations, accounts for less than 1 percent of all population growth.[24] Latin America and the Caribbean record a bit more than 8 percent of all growth, while North America (mostly the United States, but with some in Canada) accounts for under 4 percent.[25] Less than 1 percent of the world's growth in human numbers occurs in Australia, New Zealand, and the rest of Oceania.[26]

While it grows, the world's population also continues to age—meaning that our collective average age is rising.[27] (See Figure 5.) This is an inevitable result of the combination of increasing life expectancy and decreasing birth rates.[28] But population aging raises concerns among policy makers about future levels of economic growth and the fiscal soundness of programs that support the income and health care of older citizens.[29]

In 1970, the world's median age—the precise age at which half of all people are younger and half are older—was 21.5 years.[30] Overall "youthening" was the consistent trend from 1950, when the median age was 23.5 years, until 1970—a time when global fertility was high and trending higher, without compensatory increases in life expectancy.[31] Since 1970, however, with fertility falling, the median age of the world's population has risen by a bit more than two months every year.[32] By 2010 (the last year estimated), it was 28.5 years.[33] The U.N. Population Division projects the trend to begin slowing slightly later in this century, as populous generations that resulted from past rapid population growth die and are replaced by those from less populous generations born when global fertility was lower.[34]

The overall growth and aging of human population mask an unprecedented range of demographic diversity.[35] Many industrial countries are now experiencing either relatively slow population growth or absolute decline.[36] Japan, Germany, Russia, Cuba, and 17 other countries—mostly in Eastern Europe but scattered as well across Polynesia—now have fewer inhabitants with each passing year.[37] In contrast, many developing countries continue to grow rapidly and have still-large proportions of young people.[38] Median ages are nonetheless rising slowly (albeit from low bases) in many of these countries for the same reasons they did in industrial nations: increasing life expectancy and declining fertility.[39] Some developing countries already have relatively low fertility accompanied by fairly rapid aging, with China being the most often discussed example.[40]

Regionally, most of the countries growing faster than 2 percent a year are in sub-Saharan Africa (average growth rate, in a slight increase from previous U.N. estimates, of 2.5 percent), although a few are in Asia.[41] The latter continent is especially demographically diverse. Its growth rates range from a high of 7.9 percent in Oman (with large numbers of immigrants) to a negative one-tenth of 1 percent in Japan.[42] Overall, the growth rate in Asia is just over 1 percent, a bit less than that of the world as a whole.[43]

Latin America and the Caribbean is the world's most demographically homogenous major region, with a 1.1 percent growth rate, almost exactly the global average, and with few populous countries straying far from that figure.[44] Haiti is growing at 1.4 percent annually, while Guatemala grows at 2.5 percent.[45] Uruguay has population dynamics similar to the industrial world, with a 0.3 percent growth rate.[46] Relatively wealthy Chile is among the developing countries with a fertility rate below replacement, at 1.8 children per woman.[47] Mexico, once among the world's more rapidly growing countries, now expands at 1.2 percent.[48] This is somewhat above the population growth rate of the United States, to which many Mexicans historically have migrated.[49]

The industrial world also varies in its demographic dynamics, but within a narrow band of lower fertility and hence slower growth.[50] Contrary to some common perceptions, population growth continues among these wealthier countries as a whole, at almost exactly 0.3 percent a year, adding some 3.6 million people to the world annually.[51] The English-speaking countries have higher growth rates, with Australia's population expanding at 1.3 percent annually while that of the United Kingdom grows at 0.6 percent.[52] These numbers include net immigration, which in Australia and the United Kingdom, as in the United States, is a significant component of population growth.[53] Population is declining in Germany (–0.1 percent), as it is in Eastern Europe (–0.3 percent).[54] Russia's population is declining by 0.2 percent annually.[55]

One reason for Russia's loss of population until recently was a phenomenon that once was frequent among large groups of people but is now much less familiar: declining life expectancy.[56] Whether related to increasing alcoholism or social or economic stress, life expectancy at birth fell from a peak above 69 years in the late 1960s to less than 65 in the early 2000s.[57] Until life expectancy began rising in the past two or three years, Russia was among nine countries worldwide with falling life expectancies over most of the past two decades.[58] Similar causes may also characterize reduced life expectancy in Belarus and Ukraine.[59] In six countries in Africa—the Democratic Republic of the Congo, Lesotho, South Africa, Swaziland, Zambia, and Zimbabwe—civil conflict and the high prevalence of HIV/AIDS shortened lives in the past decade.[60]

On perhaps the most positive note in the new projections, U.N. Population

Division demographers believe that every country in the world is currently experiencing a longer life expectancy in the 2010–15 period than between 2000 and 2010.[61] They project continued improvement in life expectancy throughout the century, when all the new projection scenarios agree that life expectancy for the world will average 82 years, up from 70 years today.[62]

This rosy assessment of global longevity nine decades from now is perhaps the best illustration of a disconnect between demographers and the scientists who assess changing environmental conditions worldwide.[63] The U.N. demographers, like others who produce major population projections, decline to factor in the possibility that mortality trends will vary from recent history, making no mention of possible downward shifts in life expectancy due to climate change or any other environmental impacts of human activities.[64]

Indeed, the new U.N. "most likely" population projection foresees not just a long-lived human population of 10.9 billion in 2100 but one of 4.2 billion in Africa with a life expectancy of 77 years compared with 58 today.[65] Climate scientists and hydrologists, who have warned of significant decreases in food production capacity in Africa due to increases in global temperatures and growing water scarcity, might be more hesitant to say that such an outcome is "most likely" or "expected."[66] Until scientists who study global environmental change and those who assess human population dynamics find a way to reconcile their conflicting projections of the future, however, the U.N. Population Division's projections remain—for good or for ill—the crystal ball most often referred to by those wondering what humanity will look like in 2100.

Women as National Legislators

Janice Pratt and Robert Engelman

In late 2013, women accounted for slightly more than 21 percent of the representatives in the lower or popular chambers of national legislatures worldwide, according to the Geneva-based Inter-Parliamentary Union (IPU).[1] Filling one in five seats of national legislative bodies represents progress for women, but it is hardly rapid progress: 15 years ago, slightly more than 13 percent of the seats were held by women.[2] (See Figure 1.)

Low levels of female participation in parliaments undoubtedly reflect similarly low levels of participation in other political institutions as well as in social, educational, and economic spheres generally. Data on gender gaps in these areas are less uniform and authoritative. The number of women in top national executive offices—including German Chancellor Angela Merkel and Liberian President Ellen Johnson-Sirleaf—may reflect changeable political scenes in the world's 193 U.N. member states more than actual trends in women's influence in governance.

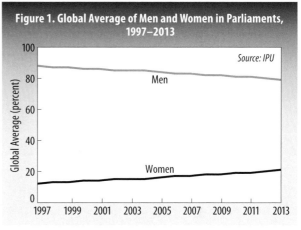

Figure 1. Global Average of Men and Women in Parliaments, 1997–2013

There are great regional variations in the average percentages of women in parliaments. (The data average upper and lower houses or express percentages for single houses in unicameral national legislatures.) As of November 2013, the figures were as follows: Nordic countries, 42 percent; Americas, 24 percent; Europe (exclusive of Nordic countries), 24 percent; sub-Saharan Africa, 22 percent; Asia, 18.5 percent; the Middle East and North Africa, 16 percent; and the Pacific, 16 percent.[3] (See Figure 2.) Where national legislatures have two houses, women tend to be better represented in the lower than the higher one—the house that in many countries has a less influential role in legislative action.

Six of the 10 parliaments with the highest number of female representation are found in developing countries, according to current data from the IPU.[4] African nations ranked impressively among the top 25 nations, with the parliaments of Rwanda, Tanzania, South Africa, the Seychelles, Angola, Uganda, and Mozambique having strong female representation.[5] Andorra, Sweden, Iceland, and Finland—the 4 industrial countries on the world's top 10 parliaments with greatest female representation—were a huge contrast to other industrial countries, including France (25 percent), the United Kingdom (22.5 percent), Greece (21 percent), and the United

Janice Pratt was an Atlas Corps Fellow from Liberia at Worldwatch Institute. **Robert Engelman** is a senior fellow at the Institute.

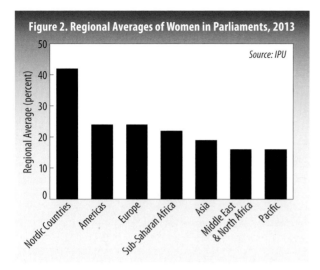

Figure 2. Regional Averages of Women in Parliaments, 2013

Source: IPU

States (18 percent), where female representation is still surprisingly low.[6]

Despite the slow progress, there is today at least the world's first example of a woman-dominated national legislature: Rwanda, with at least 56 percent representation since 2008 and 57.5 percent in 2013.[7] The post–civil war constitution of this small central African country specified that 30 percent of legislators must be women (by reserving some seats, for example, that only women can contest).[8] Rwandans are now within range of doubling that constitutionally mandated percentage, easily surpassing Cuba (49 percent) and Sweden (45 percent) in women's parliamentary representation.[9]

Although no country has managed to achieve true gender equality across social, political, and economic sectors, many have taken steps to bridge the gender gap in government, as Rwanda's constitutional effort attests. Thirty-five countries, including 9 in Africa, have so far managed to obtain 30 percent female representation in their parliaments.[10] Twenty-nine out of the 35 have in place a quota system meant to enhance women's participation in politics.[11] In India, for example, following a 1993 amendment reserving one-third of all seats in local elections for women, more than 800,000 women were elected to local village *panchayats* (a kind of village council), municipalities, and city corporations.[12] Similarly, a surge of women candidates entered Brazil's local elections in 1996 after a law required that at least 20 percent of each political party's candidates be women.[13] Other countries with some form of quota system (in some cases, simply commitments by political parties fielding candidates) include Argentina, Finland, Germany, Mexico, South Africa, and Spain.[14]

A growing body of evidence suggests that women's participation and representation in local and national governments has made a difference. A study on local councils in India found that female-led councils' initiations of drinking water projects were 62 percent higher than those of male-led councils.[15] And a study in Norway showed a positive relationship between the number of women elected to local councils and the number of childcare coverage programs implemented.[16]

Lower levels of female representation in government have limited female input into how national, regional, and local priorities are established—which excludes women's diverse and different approach to problem solving. A study conducted in the 1990s in Bolivia, Cameroon, and Malaysia indicated that if women had more say in family and community spending priorities, they would be more likely than men to improve health and education and to tackle poverty as opposed to military-related expenditures.[17]

In 2005, the World Economic Forum assessed the gender gap by measuring the extent to which women in 58 countries have achieved equality with men in political empowerment and four other critical areas: economic participation, economic opportunity, educational attainment, and health and well-being. Findings showed that Oceanic countries had the highest average overall score by region, with New Zealand ranking first in political empowerment.[18]

Although women account for more than half of the people in the world, well into the twentieth century many nations denied women the right to vote and run for office.[19] New Zealand in 1893 and Australia in 1902 were the first countries to grant electoral rights to women, but these laws applied only to women of European descent.[20] Today, only Bahrain, Kuwait, and the United Arab Emirates continue to bar women from full political participation.[21] In Kuwait, Emir Jaber al-Ahmad introduced a measure in 1999 to allow women to vote and run in elections, but the nation's all-male parliament rejected the plan.[22]

Although most countries now allow women to vote and stand for election, there is still a long way to go to achieve equal political participation. Current trends show that at the rate of growth at which women now enter parliament annually, gender equality in national legislatures may not be realized until 2068.[23]

A number of international decisions have helped legitimize the political involvement of women. Relevant treaties include the 1952 Convention on the Political Rights of Women and the Convention on the Elimination of Discrimination Against Women (CEDAW), which was adopted by the United Nations General Assembly in 1979 and entered into force two years later.[24] All but 7 of the U.N.'s 193 member countries have ratified CEDAW; the holdouts are the United States, Somalia, Sudan, South Sudan, Iran, and the two small Pacific Island nations of Palau and Tonga.[25]

In 1995, the United Nations sponsored the Fourth World Conference on Women in Beijing. With 189 governments and 2,600 nongovernmental groups in attendance, this was one of the largest U.N. conferences ever.[26] Delegates agreed to a set of strategic objectives and actions, including efforts to advance the role of women in politics and environmental stewardship.[27] The year 2015, the target year for achievement of the U.N.'s Millennium Development Goals, will also mark the twentieth anniversary of the Beijing Conference, potentially renewing attention to global efforts to empower women—not just in the world's legislative bodies but in every sphere of human activity.

Mobile Phone Growth Slows as Mobile Devices Saturate the Market

Grant Potter

More than 3.4 billion people owned at least one mobile phone in 2013.[1] This is equal to nearly half of the world's population.[2] In addition, as of 2010, more than 90 percent of people worldwide are covered by a mobile phone signal, so most people at least have access to a mobile network.[3]

The number of mobile subscriptions—that is, the number of active accounts that have access to a mobile network—far surpasses the number of phone owners. It grew from 1 billion subscriptions in 2000 to a projected figure of more than 6.8 billion by the end of 2013.[4] (See Figure 1.) This number is so much higher than the number of phone owners because many people own multiple mobile devices or multiple SIM cards for one phone.[5] As a result, the number of mobile subscriptions was expected to surpass the number of people in the world in early 2014, according to the International Telecommunication Union, a U.N. agency.[6]

However, the annual rate of growth has already begun to slow as markets become increasingly saturated.[7] Annual additions to mobile subscriptions peaked in 2010 at 680 million.[8] (See Figure 2.) The subscription rate began to dip in 2011, and it is estimated that 424 million new subscriptions were added in 2013—some 250 million fewer than in 2010.[9]

There are nearly 4 billion more active mobile phone subscriptions in the developing world than in the industrial world.[10] This is not surprising, given the distribution of world population.[11] (See Figure 3.) But on a per capita basis the picture is far different: on average, industrial countries have 128 subscriptions per 100 people, compared with 89 per 100 people in developing countries.[12] (See Figure 4.) But the figure in developing countries is expected to top 100 subscriptions per 100 people in 2014.[13]

The future of the mobile phone industry will be less about adding new subscriptions and more about improving existing service. The most common mobile network in the world uses 2G (second generation) technology that allows users to talk and send text messages. 2G accounts for nearly 4.7 billion mobile subscriptions today.[14] In the developing world, 2G is the dominant mobile platform because the network is very inexpensive to install, costing less than fixed-line networks for

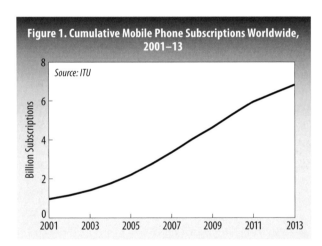

Figure 1. Cumulative Mobile Phone Subscriptions Worldwide, 2001–13

Source: ITU

Grant Potter was a development associate at the Worldwatch Institute.

wired phones.[15] The ability to set up 2G networks on difficult terrain without a lot of pre-existing infrastructure has led to "leapfrogging," in which many users skip landline technology altogether in favor of mobile phones. For example, Afghanistan had 2 million cell phone subscribers in 2010 compared with only 20,000 landline phones.[16]

As mobile networks are upgraded, 2G networks will be transitioned into 3G or 4G networks that give users access to mobile broadband Internet (colloquially known as data) or, when available, fixed wireless Internet (known as Wi-Fi). These networks now cover over 50 percent of the world population.[17] They account for approximately 75 subscriptions out of 100 in industrial countries but for only 20 subscriptions out of 100 in developing countries.[18] Although "Active mobile-broadband subscriptions" have already displaced 2G as the dominant technology in industrial countries, the reverse is true in developing countries, where 2G has nearly 3.5 times as many users.[19] (See Table 1.) This ratio is expected to shift in the next five years, however. Estimates indicate that by 2018 there will be 9.3 billion mobile subscriptions, with most of the added growth occurring in developing countries, and that mobile-broadband subscriptions will account for 6.3 billion of those devices—roughly two-thirds of the total market.[20]

Perhaps one of the most important side-effects of the growing mobile phone industry in the developing world is that financial services have become tethered to mobile phone use in poor regions. Areas with high poverty tend to have mobile subscriptions rates of 50 out of 100 people, while only 37 percent of people living there have access to a physical bank branch.[21] Financial institutions have begun to leverage the existing infrastructure for mobile phones so that a host of financial transactions—such as opening a savings account, paying bills, or transferring money—can be conducted at local mobile retail stores.[22]

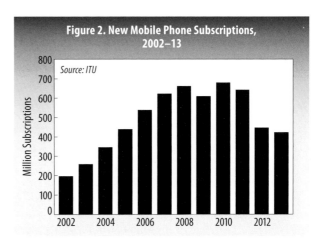

Figure 2. New Mobile Phone Subscriptions, 2002–13

Source: ITU

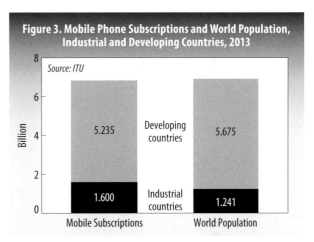

Figure 3. Mobile Phone Subscriptions and World Population, Industrial and Developing Countries, 2013

Source: ITU

Developing countries 5.235 / 5.675
Industrial countries 1.600 / 1.241

Mobile Subscriptions   World Population

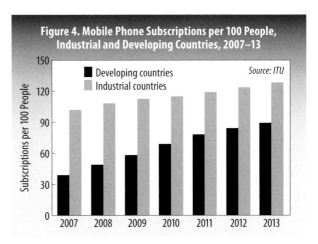

Figure 4. Mobile Phone Subscriptions per 100 People, Industrial and Developing Countries, 2007–13

Developing countries
Industrial countries

Source: ITU

Table 1. 2G and 3G/4G Networks in Industrial and Developing Countries		
Network	**Industrial Countries**	**Developing Countries**
	(percent of subscriptions)	
2G Network	42	78
3G/4G Network	58	22

Source: ITU, The World in 2013: ICT Facts and Figures (Geneva: 2013).

Mobile phones also present great opportunities for development and aid groups. Development organizations use these phones to provide information and services to their clients because people read approximately 97 percent of their text messages compared with only 5–20 percent of their e-mails, making a text message one of the most effective ways to convey important news.[23] For example, health organizations in some African countries have begun sending weekly reminders to AIDS patients to take their antiretroviral drugs instead of sending health workers out to different communities.[24] This reaches a wider audience more often and greatly trims the organizations' travel costs.[25]

Mobile phones have also helped improve farmers' incomes in developing countries. Some mobile applications, like MFarm in Kenya, provide market rates for farmers looking to sell their crops and prevent abuse from go-betweens who might capitalize on a lack of price transparency.[26] A study in Niger revealed that grain traders with mobile phones had 29 percent higher profits than traders without phones.[27]

One of the most dramatic uses of mobile phones was during the Arab Spring protests in 2010. One Egyptian activist explained, "we use Facebook to schedule the protests, Twitter to coordinate, and YouTube to tell the world."[28] The Dubai School of Government found that 9 out of 10 Egyptian and Tunisian protestors used Facebook to coordinate and popularize their protests.[29] Although social media, such as Facebook, are available on a computer, more people in Egypt own phones than computers, and mobile phones were preferred in street protests because they could be carried anywhere and easily concealed.[30] Although governments may shut down their mobile and Internet service in the face of mass protests, they do so at a high cost. When President Mubarak took Egypt offline in January 2011, it cost the country $18 million a day in lost revenue and, more important, contributed to increasing participation in anti-government demonstrations.[31]

Although mobile phones offer a lot of benefits to the developing world, their production and disposal create problems there, too. The war-torn Democratic Republic of Congo has 60–80 percent of the world's deposits of columbite-tantalite (also known as coltan), a critical component in nearly all mobile devices.[32] In addition, Congo has a wealth of other minerals used in these devices, such as gold, tantalum, and tungsten.[33] The sale of these minerals has exacerbated the deadly conflict in this country.[34] So far, efforts to reduce sales have backfired, as mine operators have partnered with armed groups to smuggle their minerals out of the country.[35]

The manufacture of mobile phones has also involved severe labor abuses. Factories get contracts from phone companies to produce the devices as cheaply as possible, and as a result they cut corners in ways that can have severe impacts on workers' health. In an investigation by China Labor Watch, labor violations in

Apple factories included overtime in excess of 150 hours a month during peak production season, working for months without a day off, assembly-line workers having to stand for 11 hours a day with minimal breaks, and poorly ventilated factories where workers breathe metal dust.[36] In 2011, as a result of improper ventilation, an explosion at a factory in Chengdu, China, that was producing iPads killed 4 people and injured 18.[37] Apple is not the only company with a problematic supply chain. Samsung's suppliers in China have been accused of excessive overtime, exhausting conditions, inhumane treatment of workers, and the hiring of underage workers.[38]

Mobile phones also create certain health and environmental problems when they are thrown away or recycled improperly. Americans replace their mobile phones once every two years on average.[39] In 2010, over 150 million phones were thrown away or recycled by Americans alone.[40] Old phones, along with other so-called e-waste, are often exported to countries like India and China, where the valuable materials contained in them are extracted in ways that endanger the health of the workers and that pollute the local environment with dangerous toxics.[41] Exposure to the phones' components can have severe neurological effects, especially on the children who are most often the ones involved in this extraction.

Notes

Introduction: A Global Disconnect (pages xi–xii)

1. Intergovernmental Panel on Climate Change, *Fifth Assessment Report of the Intergovernmental Panel on Climate Change*, three volumes (Geneva: 2013 and 2014).

2. U.S. Global Change Research Program, *National Climate Assessment: Climate Change Impacts in the United States* (Washington, DC: 2014).

3. Justin Gillis, "U.S. Climate Has Already Changed, Study Finds, Citing Heat and Floods," *New York Times*, 7 May 2014.

4. Naomi Klein, "The Change Within: The Obstacles We Face Are Not Just External," *The Nation*, 21 April 2014.

Fossil Fuels Dominate Primary Energy Consumption (pages 2–5)

1. BP, *Statistical Review of World Energy 2013* (London: 2013).

2. Ibid.

3. International Energy Agency, "Coal's Share of Global Energy Mix to Continue Rising, with Coal Closing In on Oil as World's Top Energy Source by 2017," press release (Paris: 17 December 2012).

4. Russell Gold and Daniel Gilbert, "U.S. Is Overtaking Russia as Largest Oil-and-Gas Producer," *Wall Street Journal*, 2 October 2013.

5. BP, op. cit. note 1.

6. Ibid.

7. Ibid.

8. Ibid.

9. Ibid.

10. U.S. Energy Information Administration (EIA), *International Energy Outlook 2013: Natural Gas* (Washington, DC: July 2013).

11. Ibid.

12. Ibid.

13. BP, op. cit. note 1.

14. Ibid.

15. Ibid.

16. Ibid.

17. Ibid.

18. Ibid.

19. Ibid.

20. EIA, "Multiple Factor Push Western Europe to Use Less Natural Gas and More Coal," *Today in Energy*, 27 September 2013.

21. BP, op. cit. note 1.

22. Ibid.

23. EIA, "Monthly Coal- and Natural Gas-Fired Generation Equal for First Time in April 2012," *Today in Energy*, 6 July 2012.

24. EIA, "Natural Gas Generation Lower than Last Year Because of Differences in Relative Fuel Prices," *Today in Energy*, 25 September 2013.

25. BP, op. cit. note 1.

26. Ibid.

27. Ibid.

28. Ibid.

29. Ibid.

30. Ibid.

31. Ibid.

32. Ibid.

33. Ibid.

34. Ibid.

35. Ibid.

36. Ibid.

37. Ibid.

38. Ibid.

39. Ibid.

40. Ibid.

41. Ibid.

42. Ibid.

43. Ibid.

44. Ibid.

45. EIA, *Analysis Briefs: Sudan and South Sudan*, at www.eia .gov/countries/cab.cfm?fips=SU, updated 5 September 2013.

46. Ibid.

47. Ibid.

48. Ibid.

49. Shakuntala Makhijani, "Growth in Global Oil Market Slows," in Worldwatch Institute, *Vital Signs: Volume 20* (Washington, DC: Island Press, 2013), pp. 2–5.

50. BP, op. cit. note 1.

51. Ibid.

52. Ibid.

53. Ibid.

54. EIA, *Petroleum & Other Liquids: Prices*, at www.eia.gov /dnav/pet/hist/LeafHandler.ashx?n=PET&s=RWTC&f=D, viewed 8 October 2013.

55. BP, op. cit. note 1.

56. Ibid.

57. Ibid.

58. EIA, "China Poised to Become the World's Largest Net Oil Importer Later This Year," *Today in Energy*, 9 August 2013.

Nuclear Power Recovers Slightly, But Global Future Uncertain (pages 6–9)

1. International Atomic Energy Agency (IAEA), *Nuclear Power Reactors in the World, 2013 Edition* (Vienna: 2013). Other IAEA sources report slightly different numbers.

2. Ibid.

3. "China," in IAEA, Power Reactor Information System, 2013, at www.iaea.org/pris.

4. IAEA, op, cit. note 3; Denis Langlois, "Bruce Power's Rebuilt Unit 1 Reactor Back Online," *Toronto Sun*, 21 September 2012.

5. Earth Policy Institute, "World Cumulative Installed Nuclear and Wind Power Capacity and Net Annual Additions, 1950–2008," at www.earth-policy.org/datacenter/xls /update78_3.xls; IAEA, op. cit. note 1.

6. BP, *Statistical Review of World Energy 2012*, Historical Data Excel Workbook (London: 2012).

7. IAEA, op. cit. note 3.

8. IAEA, op. cit. note 1; "United States," in IAEA, op. cit. note 3.

9. "France," in IAEA, op. cit. note 3.

10. "United States," op. cit. note 8; see other countries' pages in IAEA, op. cit. note 3, for their nuclear shares of electricity generation.

11. IAEA, op. cit. note 3.

12. "Russia," in IAEA, op. cit. note 3.

13. IAEA, op. cit. note 1; "China," op. cit. note 3.

14. "Republic of Korea," in IAEA, op. cit. note 3.

15. "Asia Airs Nuclear Ambitions at U.N. Gathering," *Reuters*, 20 September 2013; World Nuclear Association, "Nuclear Power in South Korea," updated 20 August 2013, at www.world-nuclear.org.

16. Phillip Inman and Terry Macalister, "Japan Turns Off Last Nuclear Reactor amid Fears of Surge in Gas Prices," (London) *Guardian*, 16 September 2013.

17. IAEA, op. cit. note 3.

18. Ibid.

19. IAEA, op. cit. note 1.

20. Ibid.

21. Data for 1964–2007 from Worldwatch database; 2008–11 calculated from IAEA, op. cit. note 3.

22. Osha Gray Davidson, "Germany Abandons Nuclear Power and Lives to Talk About It," *Bloomberg*, 16 November 2012.

23. "Closure Dates for Belgian Units," *World Nuclear News*, 23 July 2012.

24. Nuclear Energy Agency, *2012 NEA Annual Report* (Paris: Organisation for Economic Co-operation and Development, 2013), p. 11.

25. Ibid., p. 10.

26. Ibid., p. 11; IAEA, op. cit. note 3.

27. U.S. Energy Information Administration, "Levelized Cost of New Generation Resources in the Annual Energy Outlook 2013," January 2013, at www.eia.gov/forecasts /aeo/pdf/electricity_generation.pdf.

28. World Nuclear Association, "The Economics of Nuclear Power," updated August 2013, at www.world-nuclear.org /info/Economic-Aspects/Economics-of-Nuclear-Power.

29. For tightened safety measures, see Nuclear Energy Agency, op. cit. note 24, p. 11; for public opinion, see the example of the United States in Pew Research Center,

"As Gas Prices Pinch, Support for Oil and Gas Production Grows," 19 March 2012.

30. "Fukushima Radiation Levels '18 Times Higher' than Thought," *BBC News Asia-Pacific*, 2 September 2013; "Japan" at IAEA, op. cit. note 3; Tanya Lewis, "Fukushima Plant Springs 300-Ton Water Leak," *Live Science*, 20 August 2013.

31. World Nuclear Association, "Nuclear Power in Japan," updated 27 September 2013, at www.world-nuclear.org /info/Country-Profiles/Countries-G-N/Japan/#.UkxqPT _MDcs.

32. "Japan" op. cit. note 30; Mari Saito and Nobuhiro Kubo, "Tokyo Electric Gets OK to Seek Restart of World's Largest Nuclear Plant," *Reuters*, 26 September 2013.

33. Faith Hung and Antoni Slodkowski, "In North Asia, A Growing Crisis of Confidence in Nuclear Power," *Reuters*, 9 August 2013; "Energy Firm Says Water Leak Stopped at Nuclear Plant," *Euronews*, 6 April 2012; John Funk, "New Cracks Found in Davis-Besse Shield Building," *Cleveland Plain Dealer*, 20 September 2013; "US Nuclear Reactor Turned Off after Radiation Leak," *PhysOrg*, 2 February 2012; "Seawater Leak Shuts Down Swedish Nuclear Reactor," *RT*, 21 December 2012.

34. K. J. Kwon, "North Korea Proclaims Itself a Nuclear State in New Constitution," *CNN*, 31 May 2012.

35. "Hanford Double-Shell Tank Leaks Nuclear Waste" *Environment News Service*, 20 August 2012; Emily Yehle, "Moniz Floats 'Framework' to Jump-start Talks on Hanford Cleanup," *Greenwire*, 25 September 2013.

Growth of Global Solar and Wind Energy Continues to Outpace Other Technologies (pages 10–14)

1. BP, *Statistical Review of World Energy 2013* (London: 2013).

2. Ibid.

3. Bloomberg New Energy Finance, *Global Trends in Renewable Energy Investment in 2013* (Frankfurt: 2013).

4. REN21, *Renewables 2013 Global Status Report* (Paris: 2013).

5. Ibid.

6. Ibid.

7. Ibid.

8. Ibid.

9. Ibid.

10. BP, op. cit. note 1.

11. Ibid.

12. Ibid.

13. REN21, op. cit. note 4.

14. Ibid.

15. Ibid.

16. Mark Roca, "Italy Set to Cease Granting Tariffs for New Solar Project," *Bloomberg News*, 11 June 2013.

17. REN21, op. cit. note 4.

18. Ibid.

19. BP, op. cit. note 1.

20. Ibid.

21. Ibid.

22. Photovoltaic Power Systems Programme, *PVPS Report: A Snapshot of Global PV 1992–2012*, preliminary information from the International Energy Agency.

23. REN21, op. cit. note 4; "Feed-in Tariff for Grid-Connected Solar Power Systems," *Energy Matters*, at www.energy matters.com.au/government-rebates/feedintariff.php, viewed 10 July 2013.

24. REN21, op. cit. note 4.

25. "China to Aid Solar Industry by Easier Financing: Cabinet," *Bloomberg News*, 14 June 2013.

26. "China Releases 12th Five-Year Plan for Solar Power Development," *China Briefing*, 19 September 2012.

27. Brian Wingfield and Mark Drajem, "U.S., EU Said to Be in Talks With China to End Solar Spat," *Bloomberg News*, 14 June 2013.

28. BP, op. cit. note 1.

29. Ibid.

30. REN21, op. cit. note 4.

31. Ibid.

32. Ibid.

33. BP, op. cit. note 1.

34. Ibid.

35. REN21, op. cit. note 4.

36. Ibid.

37. BP, op. cit. note 1.

38. BP, "Wind Energy (2013)," at www.bp.com/en/global /corporate/about-bp/statistical-review-of-world-energy -2013/review-by-energy-type/renewable-energy/wind-ener gy.html.

39. REN21, op. cit. note 4.

40. Global Wind Energy Council (GWEC), "Global Installed Wind Power Capacity (2013)," at www.gwec.net/wp -content/uploads/2012/06/Global-installed-wind-power -capacity-MW-ÔÇô-Regional-Distribution.jpg.

41. Ibid.

42. GWEC and Greenpeace, *Global Wind Energy Outlook 2012* (Brussels and Amsterdam: 2012).

43. GWEC, op. cit note 40.

44. Ibid.

45. Ibid.

46. European Wind Energy Association, *Wind in Power: 2012 European Statistics* (Brussels: 2013).

47. GWEC, op. cit. note 40.

48. Ibid.

49. Ibid.

50. Ibid.

51. Ibid.

52. GWEC, *Global Wind Report: Annual Market Update 2011* (Brussels: 2012).

53. GWEC, op. cit. note 40.

54. Ibid.

55. GWEC and Greenpeace, op. cit. note 42.

56. Ibid.

57. GWEC, op. cit. note 40.

58. Ibid.

59. GWEC and Greenpeace, op. cit. note 42.

60. REN21, op. cit. note 4.

61. Ibid.

62. Ibid.

63. Ibid.

Biofuel Production Declines (pages 15–17)

1. BP, *Statistical Review of World Energy 2013* (London: 2013).

2. REN21, *Renewables 2013 Global Status Report* (Paris: 2013), p. 14.

3. BP, op. cit. note 1.

4. REN21, op. cit. note 2, p. 128.

5. Ibid., p. 125.

6. Ibid., p. 83.

7. Ibid., pp. 16, 19, 25, 28.

8. BP, op. cit. note 1, p. 17.

9. Ibid., p. 31.

10. Ibid.

11. Meghan Gordon, "US Loses Ethanol Exporter Status after Midwestern Drought," *Platts/McGraw Hill Financial*, 13 November 2012.

12. REN21, op. cit. note 2, pp. 31, 28.

13. Ibid., p. 31.

14. U.S. Department of Agriculture, Foreign Agricultural Service, *EU-27 Biofuels Annual 2013* (Washington, DC: 2013), p. 13.

15. REN21, op. cit. note 2, p. 31.

16. Ibid.

17. Ibid.

18. Ibid., p. 20.

19. Geoff Cooper, "2012 Ethanol Exports Total 739 Million Gallons, Canada Is Top Destination," *The E-Xchange* (Renewable Fuels Association), 11 February 2013.

20. Ibid.

21. Ibid.

22. U.S. Energy Information Administration (EIA), "U.S. Biomass-based Diesel Exports by Destination for 2012," at www.eia.gov/dnav/pet/pet_move_expc_a_EPOORDB _EEX_mbbl_a.htm; EIA, "U.S. Biomass-based Diesel Imports by Country of Origin for 2012," at www.eia.gov /dnav/pet/pet_move_impcus_a2_nus_EPOORDB_im0 _mbbl_a.htm.

23. EIA, "Exports," op. cit. note 22; EIA, "Imports," op. cit. note 22.

24. REN21, op. cit. note 2, p. 14.

25. Ivetta Gerasimchuk et al., *State of Play on Biofuels Subsidies: Are Policies Ready to Shift?* Global Subsidies Initiative (Winnipeg, MB: International Institute for Sustainable Development, 2012), p. 3.

26. Jim Lane, "Biofuels Mandates Around the World: 2012," *Biofuels Digest*, 22 November 2012.

27. Ibid.

28. Ibid.

29. Ibid.

30. Ibid.

31. REN21, op. cit. note 2, p. 72.

32. Zain Shauk, "EPA to Raise Biofuels Mandate," *Chron .com*, 31 January 2013.

33. Matthew Wald, "Court Overturns EPA's Biofuels Mandate," *New York Times*, 25 January 2013.

34. Union of Concerned Scientists, "Advanced Biofuel Mandates: Critical Decisions on Food vs. Fuel," fact sheet, November 2012, p. 2.

35. REN21, op. cit. note 2, p. 61.

36. The Pew Charitable Trusts, *Who's Winning the Clean Energy Race? 2012 Edition* (Washington, DC: 2012), Figure 4, p. 11.

37. Isabel Lane, "Biofuel Investment to Top $69 Billion, Says Navigant," *Biofuels Digest*, 22 July 2013.

38. Isabel Lane, "BP Scales Back EU Biofuels Efforts Amidst Policy Uncertainty," *Biofuels Digest*, 24 June 2013.

Policy Support for Renewable Energy Continues to Grow and Evolve (pages 18–22)

1. F. Zhang, *How Fit Are Feed-in Tariff Policies? Evidence from the European Wind Market*, Policy Research Working Paper 6376 (Washington, DC: World Bank, 2013).

2. REN21, *Renewables 2005 Global Status Report* (Paris: 2005); REN21, *Renewables 2013 Global Status Report* (Paris: 2013).

3. REN21, *Renewables 2005*, op. cit. note 2; REN21, *Renewables 2013*, op. cit. note 2.

4. REN21, *Renewables 2013*, op. cit. note 2.

5. BP, *Statistical Review of World Energy 2013* (London: 2013).

6. REN21, *Renewables 2013*, op. cit. note 2.

7. Ibid.

8. Ibid.

9. Ibid.

10. Ibid.

11. REN21, *Renewables 2005*, op. cit. note 2; REN21, *Renewables 2013*, op. cit. note 2.

12. REN21, *Renewables 2013*, op. cit. note 2.

13. Ibid.

14. Ibid.

15. Ibid.

16. Ibid.

17. Ibid.

18. Derived from REN21, *Renewables 2005*, op. cit. note 2, and from World Bank, "Country and Lending Groups," at data.worldbank.org/about/country-classifications/country -and-lending-groups.

19. REN21, *Renewables 2013*, op. cit. note 2.

20. Ibid.

21. Ibid.

22. Derived from REN21, *Renewables 2013*, op. cit. note 2, and from World Bank, op. cit. note 18.

23. Ibid.

24. REN21, *Renewables 2013*, op. cit. note 2.

25. Clean Energy Ministerial Secretariat, *Public-Private Roundtables at the Fourth Clean Energy Ministerial*, 17–18 April 2003, New Delhi, India (Washington, DC: June 2013).

26. REN21, *Renewables 2013*, op. cit. note 2.

27. M. Liebreich, "Bloomberg New Energy Finance Summit Keynote," presentation to the BNEF Finance Summit, 23 April 2013.

28. U.N. Environment Programme (UNEP) and Bloomberg New Energy Finance (BNEF), *Global Trends in Renewable Energy Investment 2013* (Frankfurt: 2013).

29. Zhang, op. cit. note 1.

30. REN21, *Renewables 2013*, op. cit. note 2.

31. Ibid.

32. U. Wang, "Here Comes Another Solar Trade Dispute," *Renewable Energy World*, 7 February 2013.

33. World Trade Organization (WTO), "Dispute Settlement: Dispute DS412 Canada-Certain Measures Affecting the Renewable Energy Generation Sector" at www.wto.org /english/tratop_e/dispu_e/cases_e/ds412_e.htm.

34. UNEP and BNEF, op. cit. note 28.

35. U.S. Department of Commerce, "Fact Sheet: Commerce Finds Dumping and Subsidization of Crystalline Silicon Photovoltaic Cells, Whether or Not Assembled into Modules from the People's Republic of China," 10 October 2012, at ia.ita.doc.gov/download/factsheets/factsheet_prc-solar-cells -ad-cvd-finals-20121010.pdf; B. Wingfield and W. McQuillen, "U.S. Sets Duties on Chinese, Vietnamese Wind Towers," *Bloomberg.com*, 27 July 2012.

36. W. Ma, "China Aims Tariffs on Solar-Panel Material at U.S., South Korea," *Wall Street Journal*, 19 July 2013.

37. REN21, *Renewables 2013*, op. cit. note 2.

38. International Renewable Energy Agency, *Renewable Energy Auctions in Developing Countries* (Abu Dhabi: 2013).

39. Ibid.

40. Derived from ibid. and from World Bank, op. cit. note 18.

41. M. Miller et al., *RES-E-NEXT: Next Generation of RES-E Policy Instruments* (Utrecht, Netherlands: International Energy Agency–Renewable Energy Technology Deployment, 2013).

Phasing Out Fossil Fuel Subsidies (pages 23–26)

1. International Energy Agency (IEA), *World Energy Outlook 2012* (Paris: 2012); International Monetary Fund (IMF), *Energy Subsidy Reform: Lessons and Implications* (Washington, DC: 2013)

2. IEA, op. cit. note 1, p. 69.

3. Ibid.

4. Organisation for Economic Co-operation and Development (OECD), *Inventory of Estimated Budgetary Support and Tax Expenditures for Fossil Fuels 2013* (Paris: 2013).

5. Ibid., p. 3.

6. IEA, op. cit. note 1, p. 69.

7. Ibid., p. 70.

8. Ibid.

9. Ibid.

10. OECD, op. cit. note 4, p. 38.

11. Ibid.

12. Ibid.

13. IEA, op. cit. note 1, p. 69.

14. Ibid., p. 234.

15. Ibid.

16. Oil Change International, *Low Hanging Fruit* (Washington, DC: 2012), p. 16.

17. Ibid.

18. International Institute for Sustainable Development (IISD), *Joint Submission to the UN Conference on Sustainable Development, Rio +20: A Pledge to Phase Out Fossil-fuel Subsidies* (Winnipeg, MB: 2011), p. 10.

19. Global Subsidies Initiative, *Subsidies and External Costs in Electric Power Generation: A Comparative Review of Estimates* (Winnipeg, MB: 2011), p. 24.

20. IMF, op. cit. note 1, Appendix I.

21. Ibid.

22. Ibid., p. 1.

23. G-20, "Leadership Communiqué," St. Petersburg, 2013.

24. Doug Koplow, *Phasing Out Fossil-Fuel Subsidies in the G20: A Progress Update* (Washington, DC: Oil Change International, 2012), p. 2.

25. Ibid., p. 6.

26. Shelagh Whitley, *Time to Change the Game: Fossil Fuel Subsidies and Climate* (London: Overseas Development Institute, 2013).

27. IMF, *The Unequal Benefits of Fuel Subsidies: A Review of Evidence for Developing Countries* (Washington, DC: 2010).

28. Ibid., p. 12.

29. Ibid., p. 11.

30. Ibid.

31. IEA, *Redrawing the Energy-Climate Map*, World Energy Outlook Special Report (Paris: 2013), p. 68.

32. IMF, *Case Studies on Energy Subsidy Reform: Lessons and Implications* (Washington, DC: 2013), p. 23.

33. Ibid., p. 24.

34. Ibid.

35. Ibid., p. 6.

36. Ibid., p. 22.

37. Kathy Quiano-Castro, "After Protests, Indonesia's Parliament Boosts Fuel Prices by as Much as 44%," CNN, 17 June 2013; World Bank, *Indonesia Economic Quarterly: Redirecting Spending* (Washington, DC: 2012), p. 15.

38. IEA, op. cit. note 31, p. 11.

39. Ibid.

40. Ibid., p. 9.

Record High for Global Greenhouse Gas Emissions (pages 28–31)

1. Global Carbon Project, based on T. A. Boden, G. Marland, and R. J. Andres, "Global, Regional, and National Fossil-Fuel CO_2 Emissions," Carbon Dioxide Information Analysis Center (CDIAC), Oak Ridge National Laboratory (ORNL), Oak Ridge, TN, 2013; 1959–2010 estimates for fossil fuel combustion are from CDIAC, ORNL, at cdiac.ornl.gov/trends/emis/meth_reg.html, viewed 19 November 2013. An uncertainty of ± 5 percent means that there is a 68 percent chance of emissions falling within the given range.

2. Boden, Marland, and Andres, op. cit. note 1.

3. Ibid.

4. C. Le Quéré et al., "Global Carbon Budget 2013," *Earth System Science Data Discussions*, at doi:10.5194/ess

dd-6-689-2013 and at www.earth-syst-sci-data-discuss.net/6/689/2013/essdd-6-689-2013, viewed 19 November 2013.

5. Sandrine Rastello, "Coal-Powered Financing Minimized in World Bank Energy Policy," *Bloomberg*, 27 June 2013; "U.S. Lays Out Strict Limits on Coal Funding Abroad," *Reuters*, 29 October 2013.

6. Le Quéré et al., op. cit. note 4.

7. Ibid.

8. Boden, Marland, and Andres, op. cit. note 1.

9. U.S. Environmental Protection Agency, "Overview of Greenhouse Gases," at www.epa.gov/climatechange/ghgemissions/gases.html, viewed 22 November 2013.

10. R. F. Keeling et al., "Atmospheric CO_2 Concentrations (ppm) Derived from In Situ Air Measurements at Mauna Loa, Observatory, Hawaii," Scripps Institution of Oceanography, La Jolla, CA, 8 October 2013.

11. James Hansen et al., "Target Atmosphere CO_2: Where Should Humanity Aim?" *Open Atmospheric Science Journal*, Vol. 2 (2008), pp. 217–31.

12. World Meteorological Organization (WMO), "The State of Greenhouse Gases in the Atmosphere Based on Global Observations through 2012," *WMO Greenhouse Gas Bulletin*, November 2013.

13. Le Quéré et al., op. cit. note 4.

14. WMO, op. cit. note 12.

15. W. R. Emanuel and A. C. Janetos, "Implications of Representative Concentration Pathway 4.5 Methane Emissions to Stabilize Radiative Forcing," U.S. Department of Energy, Washington, DC, 2013.

16. James Butler and Stephen Montzka, "The NOAA Annual Greenhouse Gas Index (AGGI)," NOAA Earth System Research Laboratory, Global Monitoring Division, at www.esrl.noaa.gov/gmd/aggi/aggi.html, viewed 19 November 2013.

17. E. J. Dlugokencky et al., "Conversion of NOAA Atmospheric Dry Air CH_4 Mole Fractions to a Gravimetrically-Prepared Standard Scale," *Journal of Geophysical Research*, 110 (2005), D18306.

18. CDIAC, ORNL, "Recent Greenhouse Gas Concentrations," February 2013, at cdiac.ornl.gov/pns/current_ghg.html.

19. The Fourth Assessment Report of the Intergovernmental Panel on Climate Change defines "radiative forcing" as "a measure of the influence a factor has in altering the balance of incoming and outgoing energy in the Earth-atmosphere system and is an index of the importance of the factor as a potential climate change mechanism."

20. Butler and Montzka, op. cit. note 16.

21. Ibid.

22. Ibid.

23. Le Quéré et al., op. cit. note 4.

24. Potsdam Institute for Climate Impact Research and Climate Analytics, *Turn Down the Heat: Why a 4° Warmer World Must be Avoided* (Washington, DC: World Bank, 2012).

25. Le Quéré et al., op. cit. note 4.

26. Boden, Marland, and Andres, op. cit. note 1.

27. Ibid.

28. Ibid.

29. Ibid.

30. Ibid.; 1959–2010 estimates for fossil fuel combustion are from CDIAC, op. cit. note 1.

31. Boden, Marland, and Andres, op. cit. note 1.

32. Richard Heede, "Tracing Anthropogenic Carbon Dioxide and Methane Emissions to Fossil Fuel and Cement Producers, 1854–2010," *Climate Change*, 22 November 2013.

Agriculture and Livestock Remain Major Sources of Greenhouse Gas Emissions (pages 32–35)

1. U.N. Food and Agriculture Organization (FAO), "Emissions," in *FAOSTAT Statistical Database*, at faostat.fao.org, updated 4 December 2012.

2. International Energy Agency, *CO_2 Emissions from Fuel Combustion* (Paris: 2012).

3. FAO, op. cit. note 1; Francesco Tubiello et al., "The FAOSTAT Database of Greenhouse Gas Emissions from Agriculture," *Environmental Research Letters*, 12 February 2013.

4. P. Smith et al., "Agriculture," in Intergovernmental Panel on Climate Change, *Mitigation of Climate Change, Contribution of Working Group III to the Fourth Assessment Report of the IPCC* (Cambridge, U.K.: Cambridge University Press, 2007).

5. FAO, op. cit. note 1.

6. Ibid.

7. Eugene Takle and Don Hofstrand, *Global Warming—Agriculture's Impact on Greenhouse Gas Emissions* (Ames: Iowa State University Extension and Outreach, 2008).

8. Smith et al., op. cit. note 4.

9. FAO, op. cit. note 1.

10. Ibid.

11. European Commission, *Organic Matter Decline* (Brussels: European Communities, May 2009).

12. FAO, op. cit. note 1.

13. Ibid.

14. Ibid.

15. Ibid.

16. Ibid.

17. Takle and Hofstrand, op. cit. note 7.

18. Center for Climate Solutions, "Enteric Fermentation Mitigation," at www.c2es.org/technology/factsheet/Enteric Fermentation#12, citing C. Grainger et al., "Effect of Whole Cottonseed Supplementation on Energy and Nitrogen Partitioning and Rumen Function in Dairy Cattle on a Forage and Cereal Grain Diet," *Australian Journal of Experimental Agriculture*, 2008.

19. N. Pelletier, "Neither Fish Nor Fowl—Planning Dinner Around the Carbon Intensity of Protein Sources," presented at American Association for the Advancement of Science annual meeting, 2009.

20. Smith et al., op. cit. note 4.

21. FAO, op. cit. note 1.

22. Ibid.

23. Ibid.

24. Alan Wright, *Environmental Consequences of Water Withdrawals and Drainage of Wetlands* (Gainesville, FL: Florida Cooperative Extension Service, 2009).

25. FAO, op. cit note 1.

26. Ibid.

27. Ibid.

28. Ibid.

29. Ibid.

30. Ibid.

31. FAO, "Production," in *FAOSTAT Statistical Database*, at faostat.fao.org, updated 16 January 2013.

32. Takle and Hofstrand, op. cit. note 7.

33. Smith et al., op. cit. note 4.

34. Takle and Hofstrand, op. cit. note 7.

35. Smith et al., op. cit. note 4.

36. Ibid.

37. Takle and Hofstrand, op. cit. note 7.

38. FAO, op. cit. note 1; FAO, op. cit. note 31.

39. FAO, op. cit. note 1.

40. Smith et al., op. cit. note 4.

41. Ibid.

42. FAO, op. cit. note 1.

43. Ibid.

44. Ibid.

45. Takle and Hofstrand, op. cit. note 7.

46. Smith et al., op. cit. note 4.

47. Takle and Hofstrand, op. cit. note 7.

48. Smith et al., op. cit. note 4.

Natural Catastrophes in 2012 Dominated by U.S. Weather Extremes (pages 36–39)

1. Munich Re, *Topics Geo: Natural Catastrophes 2012–Analyses, Assessments, Positions* (Munich: 2013), pp. 52–53.

2. Munich Re calculations, based on NatCatSERVICE database, 2012.

3. Munich Re, op. cit. note 1.

4. Munich Re, op. cit. note 2.

5. Munich Re, op. cit. note1.

6. Ibid.

7. Munich Re, op. cit. note 2.

8. Ibid.

9. Ibid.

10. Ibid.

11. Munich Re, op. cit. note1.

12. Munich Re, op. cit. note 2.

13. Munich Re, op. cit. note 1.

14. Ibid.

15. Ibid.

16. Ibid,

17. Ibid,

18. Munich Re, op. cit. note 2.

19. Munich Re, op. cit. note1.

20. Munich Re, op. cit. note 2

21. Ibid.

22. Ibid.

23. Ibid.

24. Ibid.

25. Ibid.

26. Munich Re, op. cit. note 1.

27. Munich Re, op. cit. note 2.

28. Ibid.

29. Ibid.

30. Ibid.

31. Ibid.

32. Ibid.

33. Ibid.

34. Ibid.

35. Ibid.

36. Munich Re, op. cit. note 1, pp. 30–37.

37. National Hurricane Center, "2012 Atlantic Hurricane Season," at www.nhc.noaa.gov/2012atlan.shtml.

38. Ibid.

39. Munich Re, op. cit. note 1.

40. Munich Re, op. cit. note 2.

41. Ibid.

42. Ibid.

43. Ibid.

44. Ibid.

45. Ibid.

46. Ibid.

47. Munich Re, op. cit. note 1.

48. Ibid., pp. 26–29.

49. Munich Re, op. cit. note 2.

50. Ibid.

51. Ibid.

52. Ibid.

53. Ibid.

54. Ibid.

55. Ibid.

56. Ibid.

57. Ibid.

58. Ibid.

59. Ibid.

60. Ibid.

61. Ibid.

62. Ibid.

63. Ibid.

64. Ibid.

Automobile Production Sets New Record, But Alternative Vehicles Grow Slowly (pages 42–45)

1. Colin Couchman, IHS Automotive, London, e-mail to author, 19 June 2013.

2. Ibid.

3. Ibid.

4. Ibid.

5. Ibid.

6. International Council on Clean Transportation, "Global Passenger Vehicle Standards Update. February 2013 Datasheet," Washington, DC.

7. Ibid.

8. Ibid.

9. European Federation for Transport and Environment, *How Clean Are Europe's Cars? An Analysis of Carmaker Progress Towards EU CO_2 Targets in 2010* (Brussels: 2011), p. 9.

10. Calculated from U.S. Environmental Protection Agency, *Light-Duty Automotive Technology, Carbon Dioxide Emissions, and Fuel Economy Trends: 1975 Through 2012* (Washington, DC: 2013), p. 22. The data are for "adjusted composite fuel economy"—covering combined city/highway laboratory drive cycles for cars and light trucks.

11. NYE Automotive Group, "Nye Automotive Uncovers the Magic of Hybrid Cars," at nyeauto.com/nye-automotive-uncovers-the-magic-of-hybrid-cars.

12. Toyota, "Hybrid Vehicle Chronology," at www.toyota-global.com/innovation/environmental_technology/hv5m.

13. Green Car Congress, "Toyota Cumulative Global Hybrid Sales Pass 5M, Nearly 2M in US," 17 April 2013, at www.greencarcongress.com/2013/04/tmc-20130417.html.

14. Green Car Congress, "Cumulative Worldwide Sales of Honda Hybrids Passes 1 Million Units," 15 October 2012, at www.greencarcongress.com/2012/10/hondahybrids-20121015.html#more.

15. Toyota, "Worldwide Sales of TMC Hybrids Top 5 Million Units," 17 April 2013, at www2.toyota.co.jp/en/news/13/04/0417.html.

16. Electric Drive Transportation Association (EDTA), "Electric Drive Sales," at electricdrive.org/index.php?ht=d /sp/i/20952/pid/20952, viewed 28 May 2013.

17. Green Car Congress, "Pike Research Forecasts Hybrids and Plug-Ins to Grow to 4% of European Market in 2020," 2 January 2013, at www.greencarcongress.com/2013/01 /pikeeuev-20130102.html#more.

18. EDTA, op. cit. note 16.

19. Ibid.

20. Clean Energy Ministerial, Electric Vehicle Initiative, and International Energy Agency, *Global EV Outlook. Understanding the Electric Vehicle Landscape to 2020* (April 2013), p. 6.

21. Ibid., p. 12.

22. Ibid.

23. Ibid., p. 6.

24. Ibid., p. 4.

25. Calculated from ibid. and from population data in Population Reference Bureau (PRB), *2012 World Population Datasheet* (Washington, DC: 2012).

26. EDTA, op. cit. note 16, p. 9; PRB, op. cit. note 25.

27. EDTA, op. cit. note 16, p. 4.

28. Ibid., p. 10.

29. Ibid., pp. 6, 10.

30. Ibid., p. 6.

31. Ibid., p. 16.

32. David Alexander and John Gartner, *Electric Vehicle Batteries. Lithium Ion Batteries for Hybrid, Plug-in Hybrid, and Battery Electric Light Duty Vehicles: Market Analysis and Forecasts*, Executive Summary (Boulder, CO, and Washington, DC: Navigant Research, January 2013).

33. SignumBOX, "Lithium Industry: Outlook and Perspectives," 23 April 2013, at globalxfunds.com/commodities /lithiumetf/whitepaper/Lithium Presentation 06-19-12.pdf.

34. Ibid.

35. Ibid.

36. Ibid.

37. Alexander and Gartner, op. cit. note 32.

38. Andreas Dinger et al., "Batteries in Electric Cars, Challenges, Opportunities, and the Outlook in 2020," Boston Consulting Group, January 2010.

39. Marcy Lowe et al., *Lithium Ion Batteries for Electric Vehicles: The U.S. Value Chain* (Durham, NC: Center on Global-ization Governance & Competitiveness, Duke University, 2010).

40. Wolfgang Bernhart et al., *E-mobility Index for Q1 2013* (Roland Berger Strategy Consultants and Automotive Competence Center & Forschungsgesellschaft Kraftfahrwesen Aachen, 2013).

41. "U.S. Electric Car Policy to Cost $7.5 Billion by 2019: CBO," *Reuters*, 20 September 2012.

42. "U.S. Backs Off Goal of One Million Electric Cars by 2015," *Reuters*, 31 January 2013.

Air Transport Keeps Expanding (pages 46–49)

1. International Civil Aviation Organization (ICAO), *Annual Report of the Council 2012* (Montreal: 2013), Appendix 1. Figure 1 is based on data from this report and from earlier editions.

2. Calculated from ibid. and from Population Reference Bureau, *2012 World Population Data Sheet* (Washington, DC: 2012).

3. ICAO, op. cit. note 1.

4. Calculation by author based on data in ibid.

5. Ibid.

6. Ibid.

7. Ibid.

8. Ibid.

9. Ibid.

10. Ibid.

11. Ibid.

12. Ibid. Excludes aircraft with a maximum takeoff mass of less than 9,000 kilograms.

13. Boeing, *Current Market Outlook 2013–2032* (Seattle: 2013), p. 7. Note that Boeing offers a different figure for the global aircraft fleet (20,310 at the end of 2012) than the ICAO. The difference is mostly explained by the fact that ICAO includes turboprop planes, whereas Boeing focuses on jets only.

14. Boeing, op. cit. note 13.

15. Ibid., p. 15; Airbus, *Future Journeys. Global Market Forecast 2013–2032* (Blagnanc, France: 2013).

16. Flightglobal, "World Airline Rankings: Regional Picture," 22 July 2011, at www.flightglobal.com/news/articles /world-airline-rankings-regional-picture-359695.

17. Jens Flottau and Cathy Buyck, "Airline Alliances Face New Cooperative Forces," *Aviation Week & Space Technology*, 29 April 2013; Joe Sharkey, "Forget the Airline's Name; It's

All About Alliances," *New York Times*, 5 December 2011. Table 2 is based on Star Alliance, "Facts & Figures," at www.staralliance.com/en/about, on SkyTeam, "Skyteam Airline Member Benefits," at www.skyteam.com/Global/Press/Facts%20and%20figures/2013%20Feb%20update/20130215_Member%20Benefits%20Fact%20Sheet.pdf, and on Oneworld, "Oneworld at a Glance," 24 October 2013, at www.oneworld.com/news-information/oneworld-fact-sheets/oneworld-at-a-glance.

18. Flottau and Buyck, op. cit. note 17.

19. Airbus, "Orders and Deliveries," at www.airbus.com/company/market/orders-deliveries; Boeing, "Orders and Deliveries," at active.boeing.com/commercial/orders/index.cfm.

20. Calculated from Airbus, op. cit. note 19, and from Boeing, op. cit. note 19.

21. Glennon J. Harrison, "Challenge to the Boeing-Airbus Duopoly in Civil Aircraft: Issues for Competitiveness," Congressional Research Service, 25 July 2011.

22. Boeing, op. cit. note 13, p. 16.

23. Central Intelligence Agency (CIA), "Field Listing: Airports," *The World Factbook*, at www.cia.gov/library/publications/the-world-factbook/fields/2053.html.

24. Ibid.

25. Ibid.

26. ICAO, op. cit. note 1. Figure of 77 million aircraft movements is for 2011, from Airports Council International (ACI), "ACI Releases Its 2011 World Airport Traffic Report: Airport Passenger Traffic Remains Strong as Cargo Traffic Weakens," press release (Montreal: 27 August 2012).

27. ICAO, op. cit. note 1.

28. Ibid.

29. ACI, "Passenger Traffic 2000 FINAL," 1 November 2001, at www.aci.aero/Data-Centre/Annual-Traffic-Data/Passengers/2000-final.

30. ACI, "International Passenger Traffic for Past 12 Months, 12-MONTHS ENDING AUG 2013," 18 November 2013, at www.aci.aero/Data-Centre/Monthly-Traffic-Data/International-Passenger-Rankings/12-months.

31. CIA, op. cit. note 23.

32. ACI, "Cargo Traffic 2000 FINAL," 1 November 2001, at www.aci.aero/Data-Centre/Annual-Traffic-Data/Cargo/2000-final.

33. International Council on Clean Transportation (ICCT), "Programs / Aviation," at www.theicct.org/aviation.

34. Ibid.

35. Ibid.

36. Daniel Rutherford and Mazyar Zeinali, *Efficiency Trends for New Commercial Jet Aircraft, 1960 to 2008* (Washington, DC: ICCT, 2009).

37. U.S. Energy Information Administration, "Spot Prices," 27 November 2013, at www.eia.gov/dnav/pet/pet_pri_spt_s1_m.htm.

38. Ibid.

39. ICCT, "International Civil Aviation Organization's CO_2 Certification Requirement for New Aircraft," August 2013, at www.theicct.org/sites/default/files/publications/ICCTupdate_ICAO_CO2cert_aug2013a.pdf.

40. Mazyar Zeinali et al., *U.S. Domestic Airline Fuel Efficiency Ranking 2010* (Washington, DC: ICCT, 2013), pp. 2–3.

Agricultural Population Growth Marginal as Nonagricultural Population Soars (pages 52–55)

1. U.N. Food and Agriculture Organization (FAO), "Annual Population Data," FAOSTAT, at faostat3.fao.org/faostat-gateway/go/to/download/O/OA/E; definition from FAO, "Glossary," FAOSTAT, at faostat.fao.org/site/375/default.aspx.

2. FAO, "Annual Population Data," op. cit. note 1.

3. Ibid.

4. Ibid.

5. Ibid.

6. Ibid.

7. Ibid.

8. Ibid.

9. Ibid.

10. Ibid.

11. Ibid.

12. Ibid.

13. Ibid.

14. Ibid.

15. Ibid.

16. U.S. Department of Agriculture (USDA), "2007 Census of Agriculture: Farm Numbers," at www.agcensus.usda.gov/Publications/2007/Online_Highlights/Fact_Sheets/Farm_Numbers/farm_numbers.pdf.

17. FAO, "Annual Population Data," op. cit. note 1.

18. FAO, "Production Indices," FAOSTAT, at faostat3.fao.org/faostat-gateway/go/to/download/Q/QI/E.

19. Ibid.

20. Ibid.

21. Ibid.

22. FAO, "Resources," FAOSTAT, at faostat3.fao.org/faostat -gateway/go/to/download/R/*/E.

23. Arthur Grube et al., *Pesticides Industry Sales and Usage, 2006 and 2007 Market Estimates* (Washington, DC: U.S. Environmental Protection Agency, 2011).

24. Keith Fuglie, James MacDonald, and Eldon Ball, "Productivity Growth in U.S. Agriculture," Economic Brief No. EB-9 (Washington, DC: USDA, Economic Research Service, 2007).

25. FAO, "Annual Population Data," op. cit. note 1.

26. FAO, "Global Agriculture Towards 2050," How to Feed the World 2050: High-Level Expert Forum, Rome, 12–13 October 2009.

27. FAO, *Global Food Losses and Food Waste* (Rome: 2011).

Global Food Prices Continue to Rise (pages 56–60)

1. World Bank, Global Economic Monitor (GEM) Commodities Database, at databank.worldbank.org/data /views/variableselection/selectvariables.aspx?source=global -economic-monitor-commodities, viewed 11 March 2013. Unless otherwise indicated, all discussions of price are in real terms (i.e., adjusted for inflation) and refer to consumer prices, rather than prices paid to producers.

2. Ibid.

3. Ibid.

4. U.N. Population Division, *World Population Prospects: The 2010 Revision*, CD-ROM Edition (New York: 2011); U.N. Food and Agriculture Organization (FAO),"Net Per Capita Production Index," FAOSTAT, at www.faostat.fao.org /site/612/default.aspx#ancor, viewed 12 March 2013.

5. FAO, "Agricultural Area Data," FAOSTAT, at faostat.fao .org/site/377/default.aspx#ancor, viewed 12 March 2013.

6. FAO, "FAO Food Price Index," www.fao.org/worldfood situation/wfs-home/foodpricesindex/en/, viewed 11 March 2013.

7. World Bank, op. cit. note 1.

8. FAO, op. cit. note 6.

9. Ibid.

10. Ibid.

11. FAO, *Global Food Price Monitor* (Rome: 11 February 2013).

12. FAO, "Rice Market Monitor (RMM)," www.fao.org /economic/est/publications/rice-publications/rice-market -monitor-rmm/en/, viewed 13 March 2013.

13. World Bank, *Food Price Watch* (Washington, DC: November 2012).

14. World Bank, "Food Price Hike Drives 44 Million People Into Poverty," press release (Washington, DC: 15 February 2011); FAO, op. cit. note 6.

15. FAO, op. cit. note 6.

16. Ibid.

17. FAO, *Food Outlook* (Rome: November 2012), pp. 49–50.

18. Ibid., p. 50.

19. FAO, op. cit. note 6.

20. Ibid.

21. Ibid.

22. FAO, op. cit. note 17, p. 57.

23. U.N. Population Division, op. cit. note 4.

24. International Labour Organization, *Global Wage Report 2012/13: Wages and Equitable Growth* (Geneva: 2013), p. xiii.

25. FAO, "Food Supply, Crops Primary Equivalent," FAOSTAT, at faostat.fao.org/site/609/default.aspx#ancor, viewed 14 March 2013.

26. Ronald Trostle, *Global Agricultural Supply and Demand: Factors Contributing to the Recent Increase in Food Commodity Prices* (Washington, DC: U.S. Department of Agriculture Economic Research Service, May 2008), pp. 13–19.

27. U.S. Energy Information Administration, "International Energy Statistics: Biofuels Production," at www.eia.gov /cfapps/ipdbproject/iedindex3.cfm?tid=79&pid=79&aid =1&cid=ww,&syid=2000&eyid=2011&unit=TBPD, viewed 19 March 2013.

28. Trostle, op. cit. note 26.

29. Ibid.

30. Joachim von Braun and Getaw Tadesse, *Global Food Price Volatility and Spikes: An Overview of Costs, Causes, and Solutions* (Bonn: University of Bonn Center for Development Research, 2012), p. 19.

31. Ibid., p. 23.

32. Joachim von Braun and Maximo Torero, *Physical and Virtual Global Food Reserves to Protect the Poor and Prevent Market Failure*, Policy Brief 4 (Washington, DC: International Food Policy Research Institute, 2008).

33. Ramesh Sharma, *Food Export Restrictions: Review of the 2007–2010 Experience and Considerations for Disciplining*

Restrictive Measures, FAO Commodity and Trade Policy Research Working Paper No. 32 (Rome: 1 May 2011).

34. Ibid.

35. von Braun and Tadesse, op. cit. note 30.

36. Eugenio S. Bobenrieth H. and Brian D. Wright, *The Food Price Crisis of 2007/2008: Evidence and Implications* (Concepcion, Chile, and Berkeley, CA: University of Concepcion and University of California at Berkeley, 2009), pp. 9–10.

37. Ibid.

38. World Bank, op. cit. note 1.

39. Oxfam International, "Key Eurozone Banks Step Back from Food Speculation," press release (Oxford, U.K.: 18 February 2013).

Agricultural Subsidies Remain a Staple in the Industrial World (pages 61–64)

1. Organisation for Economic Co-operation and Development (OECD), *Agricultural Policy Monitoring and Evaluation 2013: OECD Countries and Emerging Economies* (Paris: 2013), pp. 302–03.

2. Ibid., p. 15

3. Ibid., p. 79.

4. Ibid.

5. Ibid.

6. Ibid.

7. Ibid.

8. Ibid., pp. 302–03.

9. Ibid.

10. Ibid.

11. European Commission, "CAP Expenditure in the Total EU Expenditure," February 2014, at ec.europa.eu/agriculture/cap-post-2013/graphs/graph1_en.pdf.

12. OECD, op. cit. note 1, pp. 302–03.

13. Ibid.

14. Ibid.

15. Chris Edwards, "Downsizing the Federal Government: Agricultural Subsidies," *The CATO Institute*, June 2009, at www.downsizinggovernment.org/agriculture/subsidies.

16. European Commission, *Agriculture and Rural Development: Direct payments*, at ec.europa.eu/agriculture/direct-support/direct-payments/index_en.htm.

17. Ron Nixon, "Senate Passes Long-Stalled Farm Bill, With Clear Winners and Losers," *New York Times*, 4 February 2014.

18. Ibid.

19. Allison Aubrey, "Congress Poised to Make Crop Insurance Subsidies More Generous," *National Public Radio*, 30 May 2013.

20. Bruce Babcock, *Taxpayers, Crop Insurance, and the Drought of 2012* (Washington, DC: Environmental Working Group (EWG), 2013), p. 4.

21. Ibid., p. 5.

22. Bruce Babcock, *Cutting Waste in the Crop Insurance Program* (Washington, DC: EWG, 2013), p. 10.

23. George Monbiot, "Farming Subsidies: This is the Most Blatant Transfer of Cash to the Rich," (London) *Guardian*, 1 July 2013.

24. "Rich Landowners Paid Millions in Farming Subsidies," *BBC*, 5 March 2012.

25. EWG, "Government Records Show Crop Insurance Subsidies Are a Boon to Big Farm Interests," press release (Washington, DC: 31 May 2012).

26. Ibid.

27. Laura Reynolds and Danielle Nierenberg, *Innovations in Sustainable Agriculture*, Worldwatch Report 188 (Washington, DC: Worldwatch Institute, December 2012), p. 9.

28. Tom Philpott, "Why This Year's Gulf Dead Zone Is Twice As Big As Last Year's," *Mother Jones*, August 2013.

29. Robert L. Thompson, "Agricultural Price Supports," *Concise Encyclopedia of Economics*, at www.econlib.org/library/Enc1/AgriculturalPriceSupports.html.

30. Ibid.

31. OECD, op. cit. note 1, pp. 123–24.

32. Ibid., p. 125.

33. Ibid.

34. Thompson, op. cit. note 29.

35. Mark Tran, "EU Agriculture Policy 'Still Hurting Farmers in Developing Countries'," (London) *Guardian*, 11 October 2011.

36. World Trade Organization (WTO), "Export Subsidies and Competition," 1 December 2004, at www.wto.org/english/tratop_e/agric_e/negs_bkgrnd08_export_e.htm.

37. International Monetary Fund, "Global Trade and the Developing Countries," Issue Brief, Washington, DC, November 2001.

38. U.N. Conference on Trade and Development, "Sustain-

able Agriculture and Food Security in LDCs," UNCTAD Policy Brief No. 20, Geneva, May 2011.

39. Ibid.

40. Emmanuel Asmah and Brandon Routman, "Removing Barriers to Improve the Competitiveness of Africa's Agriculture," in Brookings Institution, *Improving AGOA: Toward a New Framework for U.S.-Africa Commercial Engagement* (Washington, DC: 2011), pp. 16–17.

41. Ibid.

42. Charles E. Hanrahan and Randy Schnepf, *WTO Doha Round: The Agricultural Negotiations*, CRS Report for Congress (Washington, DC: Congressional Research Service, updated 22 January 2007).

43. WTO, "Days 3, 4 and 5: Round-the-Clock Consultations Produce 'Bali Package'," press release (Geneva: 5–7 December 2013).

Global Economy: Looks Good from Afar But Is Far from Good (pages 66–69)

1. International Monetary Fund (IMF), *World Economic Outlook 2013* (Washington, DC: 2013).

2. Ibid.

3. IMF, "World Economic Outlook Database," October 2013, at www.imf.org/external/data.htm.

4. Paul Krugman and Maurice Obstfeld, *International Economics: Theory & Policy* (New York: Addison-Wesley, 2009).

5. Tim Callen, "PPP vs. the Market: Which Weight Matters," *Finance and Development,* March 2007; Krugman and Obstfeld, op. cit. note 4.

6. IMF, op. cit. note 1.

7. Ibid.

8. Ibid.

9. Ibid.

10. IMF, op. cit. note 3.

11. Mark Konold, "Commodities Supercycle Slows Down in 2012," *Vital Signs Online,* Worldwatch Institute, 27 September 2013.

12. Ibid.

13. International Labour Organization, *Global Wage Report 2012/13* (Geneva: 2013).

14. Ibid.

15. Ibid.

16. Ibid.

17. Ibid.

18. Benjamin Bridgman et al., "Accounting for Household Production in the National Accounts, 1965–2010," *Survey of Current Business*, May 2012.

19. Ibid.

20. Ida Kubiszewski et al., "Beyond GDP: Measuring and Achieving Global Genuine Progress," *Ecological Economics*, vol. 93 (2013), pp. 57–68.

21. Ibid.

22. Ibid.

23. John Helliwell, Richard Layard, and Jeffrey Sachs, eds., *World Happiness Report 2013* (New York: Sustainable Development Solutions Network et al., 2013), p. 3.

24. Ibid.

25. Ibid.

26. Ibid.

27. Ibid.

28. Ibid.

29. Ibid.

30. U.N. Development Programme, *Human Development Report* (New York: various years).

31. Jennie Moore and William E Rees, "Getting to One-Planet Living," in Worldwatch Institute, *State of the World 2013* (Washington, DC: Island Press, 2013), pp. 39–50.

32. Global Footprint Network, "Earth Overshoot Day," at www.footprintnetwork.org/en/index.php/GFN/page/earth_overshoot_day.

33. Moore and Rees, op. cit. 31.

More Businesses Pursue Triple Bottom Line for a Sustainable Economy (pages 70-75)

1. For a brief history of the term triple bottom line, see "Triple Bottom Line," *The Economist*, 17 November 2009.

2. Global Reporting Initiative (GRI), *Sustainability Disclosure Database*, at database.globalreporting.org; GRI, *The GRI Reports List 1999–2013*, at www.globalreporting.org/resourcelibrary/GRI-Reports-List-1999-2013.zip.

3. B Lab and Green America, discussions with author.

4. B Lab staff, discussions with author, April 2013; press reports in states where statutes have been enacted.

5. William H. Clark, Jr., and Larry Vranka, *White Paper: The Need and Rationale for the Benefit Corporation: Why It Is the Legal Form that Best Addresses the Needs of Social Entrepreneurs, Investors, and, Ultimately, the Public*, 18 January 2013, Benefit Corp Information Center, at www.benefitcorp.net.

6. Ibid.

7. See ibid. for proponents' arguments on why a new corporate status of benefit corporation is legally needed and the best alternative.

8. Independent Sector, "Benefit Corporations," at www.independentsector.org/benefit_corporations.

9. B Lab, "The Non-Profit Behind B Corps," at www.bcorporation.net.

10. Benefit Corp Information Center, "State by State Legislative Status," at www.benefitcorp.net.

11. A recent B Lab statement reporting that the state bar in Delaware had drafted the proposed legislation to establish benefit corporation status in that state characterized it as signaling "a seismic shift in corporate law," since Delaware is the legal home for more than 50 percent of all public companies and two thirds of all Fortune 500 companies. See www.benefitcorp.net/storage/documents/Delaware_Benefit_Corporation_Legislation.pdf.

12. Certified B Corporation, or B Corp, is a private certification by B Lab—not to be confused with a company that elects to legally register, with its incorporating documents, with the new legal status of a benefit corporation.

13. Heather R. Van Dusen, senior associate, B Lab, discussions with author, early April 2013.

14. B Lab staff, discussions with author, early April 2013.

15. Hugo Martin, "Outdoor Retailer Patagonia Puts Environment Ahead of Sales Growth," *Los Angeles Times*, 24 May 2012; Dan D'Ambrosio, "King Arthur Flour to Begin Expansion in June," *Burlington Free Press*, 10 May 2011.

16. B Lab staff, op. cit. note 14.

17. Ibid.

18. Ibid. By early April 2013, there were 582 Certified B Corporations in the United States, 66 in Canada, 23 in Chile, 8 in Argentina, 7 in Colombia, 6 in Mexico, 5 in Australia, 4 in Brazil and India, 2 each in the United Kingdom, New Zealand, and Kenya, and 1 each in Afghanistan, Belgium, Costa Rica, Germany, Guatemala, China (Hong Kong), Ireland, Italy, Republic of Korea, Mongolia, Nicaragua, Peru, Tanzania, and Turkey.

19. B Lab staff, op. cit. note 14.

20. For state-by-state details of legal requirements for B Corp certification, see www.bcorporation.net/become-a-b-corp/how-to-become-a-b-corp/legal-roadmap/corporation-legal-roadmap.

21. B Lab staff, op. cit. note 14.

22. American Sustainable Business Council, at www.asbcouncil.org.

23. Todd Larsen, division director of Corporate Responsibility Programs, Green America, e-mail to and discussions with author, April 2013.

24. Ibid.; see also www.greenbusinessnetwork.org/for-members-/benefit-corporations.html.

25. Seventh Generation, a privately held company, does not release its annual financial data, but a 2012 report on the 50 household and personal products companies with the highest annual gross sales ranked Seventh Generation as thirty-eighth, with 2011 sales listed at $165 million; "The Year That Was . . . Just Wasn't Very Good," *HAPPI Magazine*, July 2012.

26. From a summary of Green America's history in certifying green businesses, according to Larsen, op. cit. note 23. (Co-op America became Green America in 2009.)

27. Marjella Alma, external relations manager for the Global Reporting Initiative's Focal Point USA, discussions with author, April 2013. See also the Global Initiative for Sustainability Ratings at ratesustainability.org.

28. Larsen, op. cit. note 23.

29. Ibid.

30. Ibid.

31. Since the shift in process, Green America has certified 374 companies, with only 36 earning a Gold level certification, 179 being certified at the Silver level, and 159 at the Bronze level. About 3,000 businesses were awarded the Gold level based on the previous screening process.

32. Larsen, op. cit. note 23.

33. Ibid.

34. Van Dusen, op. cit. note 13.

35. Ibid.

36. Data provided by B Lab, April 2013.

37. Ibid.

38. Based on an analysis of GRI, *The GRI Reports List 1999–2013*, op. cit. note 2.

39. Ibid.

40. Global Sustainable Investment Alliance, *Global Sustainable Investment Review 2012* (2013), p. 2.

41. Ibid.

42. Ibid.

43. Ibid.

Development Aid Falls Short, While Other Financial Flows Show Rising Volatility (pages 76–79)

1. The Organisation for Economic Co-operation and De-

velopment (OECD) defines official development assistance (ODA) as financial assistance that meets three key criteria: administered by the public (or official) sector, provided with economic development as the primary objective, and given below market rates; OECD, "Net ODA from DAC Countries from 1950 to 2012, April 2013," at www.oecd.org/dac/stats/data.htm.

2. OECD, op. cit. note 1.

3. Ibid.

4. OECD, "Preliminary Data–ODA Data for 2012," April 2013, at www.oecd.org/dac/stats/data.htm.

5. Ibid.

6. Gross national income refers to total gross domestic product plus net income from foreign sources.

7. OECD, op. cit. note 4.

8. Ibid.

9. Ibid.

10. U.N. General Assembly, "International Financial System and Development: Report of the Secretary-General," New York, 27 July 2012.

11. Ibid.

12. Development Initiatives, *Global Humanitarian Assistance Report 2013* (Geneva: 2013).

13. Ibid.

14. Ibid.

15. Ibid.

16. Ibid.

17. Ibid.

18. Ibid.

19. International Monetary Fund (IMF), World Economic Outlook Database, April 2013, at www.imf.org/external/pubs/ft/weo/2013/01/weodata/download.aspx.

20. Ibid.

21. Ibid.

22. Ibid.

23. Ibid.

24. Ibid.

25. Ibid.

26. Ibid.

27. Ibid.

28. United Nations, *World Economic Situation and Prospects*

2013 (New York: 2013), p. 70.

29. Ibid.

30. Ibid.

31. United Nations Conference on Trade and Development, *World Investment Report 2013* (Geneva: 2013), p. 40.

32. "As China Prepares to Ramp Up African Investment by $2.4B, Obama Shrugs Off Competition," *International Business Times*, 1 July 2013.

33. IMF, op. cit. note 19.

34. Ibid.

35. United Nations, *World Economic Situation and Prospects 2012* (New York: 2012), p. 67.

36. Ibid.

37. Ibid., p. 68.

38. Ibid., p. 70.

39. U.N. General Assembly, op. cit. note 10.

Commodity Supercycle Slows Down in 2012 (pages 80–83)

1. Calculations based on World Bank, Commodity Price Data, September 2013. Prices for annual indices used in these calculations are real value adjusted using the World Bank's Manufacturing Unit Value Index, with 2005 = 100.

2. Javier Blas, "Supercycle Runs Out of Steam—For Now," *Financial Times*, 17 July 2012.

3. World Bank, op. cit. note 1.

4. Blas, op. cit. note 2.

5. U.N. Conference on Trade and Development, *Global Commodities Forum: Harnessing Development Gains from Commodities Production and Trade* (Geneva: 2012), p. 9.

6. Pauline Skypala, "Enthusiasm Wanes As Regulators Sharpen Focus," *Financial Times*, 1 December 2012.

7. Ibid.

8. Ibid.

9. International Monetary Fund (IMF), *World Economic Outlook* (Washington, DC: April 2013), p. 27. Figure 1 is an average of West Texas Intermediate and Brent prices.

10. Ibid., p. 28.

11. Ibid.

12. Ibid.

13. Ibid., p. 27.

14. Ibid., p. 28.

15. Ibid., p. 29.

16. U.S. Department of Agriculture (USDA), *Crop Production 2012 Summary* (Washington, DC: January 2013).

17. Gary Vocke and Olga Liefert, "Wheat Outlook: Lower Exports Raise Ending Stocks," USDA, Washington, DC, December 2012.

18. Ronald Trostle, "Global Agricultural Supply and Demand: Factors Contributing to the Recent Increase in Food Commodity Prices," USDA, Washington, DC, May 2008, p. 13.

19. L. Plantier, *Commodity Markets and Commodity Mutual Funds*, ICI Research Perspective 18, no. 3 (Washington, DC: Investment Company Institute, May 2012), p. 9.

20. IMF, op. cit. note 9, p. 28.

21. Ibid.

22. Philip C. Abbott, Christopher Hurt, and Wallace E. Tyner, *What's Driving Food Prices in 2011?* (Oak Brook, IL: Farm Foundation, NFP, 2011), p. 2.

23. Ibid., p. 3.

24. Minefund, "Snapshot of Precious Metals Price Performance in 2012," at minefund.com/wordpress/snapshot-of-precious-metals-price-performance-in-2012.

25. World Gold Council, *Gold Demand Trends: Full Year 2012* (London: December 2012), p. 7.

26. Minefund, op. cit note 24 .

27. John Bagges and Damir Ćosić, *Global Economic Prospects: Commodity Markets Outlook* (Washington, DC: World Bank, July 2013), p. 8.

28. Ibid.

29. Ibid.

30. Jack Farchy, "Storage Stacks Up for Traders," *Financial Times*, 17 July 2012.

31. Ibid.

32. Jack Farchy, "Investors Flee from Commodities at Record Pace," *Financial Times*, 19 July 2013.

Military Expenditures Remain Near Peak (pages 86–89)

1. Unless otherwise noted, all monetary terms are expressed in constant 2011 dollars. Stockholm International Peace Research Institute (SIPRI), SIPRI Military Expenditure Database, at www.sipri.org/research/armaments/milex/milex_database.

2. Ibid.

3. Ibid.

4. Ibid.

5. Susan T. Jackson, "Key Developments in the Main Arms-Producing Countries, 2011–12," in SIPRI, *SIPRI Yearbook 2013: Armaments, Disarmament and International Security* (New York: Oxford University Press, 2013), p. 206; Figure 2 compiled from ibid. and from earlier *SIPRI Yearbooks* (2005 to 2012 editions).

6. Ibid., p. 228.

7. SIPRI, op. cit. note 1.

8. Ibid.

9. Ibid.

10. Ibid.

11. Ibid.

12. Sam Perlo-Freeman, Carina Solmirano, and Helén Wilandh, "Global Developments in Military Expenditure," in SIPRI, op. cit. note 5, p. 131.

13. Ibid.

14. Ibid., p. 133.

15. Calculated from SIPRI, op. cit. note 1.

16. Ibid.

17. Calculated from ibid.

18. U.S. Department of Defense, Office of the Underscretary of Defense (Comptroller), *National Defense Budget Estimates for FY 2014* (Washington, DC: May 2013). This publication reports budget figures in constant 2005 dollars, which have been translated into constant 2011 dollars to make them more comparable with the SIPRI data.

19. U.S. Department of Defense, Defense Manpower Data Center (DMDC), "Active Duty Military Personnel by Service by Region/Country," as of 31 July 2013, at www.dmdc.osd.mil/appj/dwp/reports.do?category=reports&subCat=milActDutReg.

20. Ibid.

21. SIPRI, op. cit. note 1.

22. Ibid.

23. Ibid.

24. A list of various global social problems, with estimates for the number of people affected, has been compiled by Anup Shah, "Poverty Facts and Stats," *Global Issues*, last updated 7 January 2013, at www.globalissues.org/article/26/poverty-facts-and-stats.

25. World Bank, "Poverty Overview," at www.worldbank.org/en/topic/poverty/overview; Perlo-Freeman, Solmirano, and Wilandh, op. cit. note 12, p. 125.

26. High-income countries' military spending from Perlo-Freeman, Solmirano, and Wilandh, op. cit. note 12, p. 129; development assistance data from Organisation for Economic Co-operation and Development, "Net ODA from DAC Countries from 1950 to 2012, April 2013," at www.oecd.org/dac/stats/data.htm.

27. Friends Committee on National Legislation, "Where Do Our Income Tax Dollars Go?" at www.fcnl.org/assets/flyer/FCNL_Taxes12.pdf.

28. Ibid.

29. Scott Shane, "New Leaked Document Outlines U.S. Spending on Intelligence Agencies," *New York Times*, 29 August 2013; comparison to countries' military budgets from SIPRI, op. cit. note 1.

30. Wilson Andrews and Todd Lindeman, "$52.6 Billion: The Black Budget," *Washington Post*, 29 August 2013.

31. Dana Priest and William M. Arkin, "A Hidden World, Growing Beyond Control," *Washington Post*, 19 July 2010.

Peacekeeping Budgets Equal Less Than Two Days of Military Spending (pages 90–94)

1. U.N. Department of Public Information (UNDPI), "United Nations Peacekeeping Operations. Fact Sheet: 31 January 2014" (New York: February 2014). UNDPI reports budgets in current dollars; dollar amounts here are in 2013 dollars.

2. Ibid.; UNDPI, "United Nations Peacekeeping Operations. Fact Sheets," earlier editions (prior to January 2011, the Fact Sheets were called Background Notes); Worldwatch database.

3. Cumulative peacekeeping in 2013 dollars spending calculated from "United Nations Peacekeeping Operations. Fact Sheets," and from Worldwatch database. In current dollar terms, cumulative budgets came to $99 billion. Military expenditures from Stockholm International Peace Research Institute (SIPRI), SIPRI Military Expenditure Database, at www.sipri.org/research/armaments/milex/milex_database.

4. UNDPI, op. cit. note 1.

5. Uniformed personnel from U.N. Department of Peacekeeping Operations (UNDPKO), "Troop and Police Contributors," at www.un.org/en/peacekeeping/resources/statistics/contributors.shtml; civilian personnel from UNDPI, op. cit. note 1.

6. GlobalSecurity.org, "World's Largest Armies," at www.globalsecurity.org/military/world/armies.htm, viewed 27 March 2014.

7. Michael Renner, "Peacekeeping Budgets and Personnel Soar to New Heights," in Worldwatch Institute, *Vital Signs 2009* (Washington, DC: 2009), pp. 80–82.

8. UNDPKO, "Troop and Police Contributors Archive (1990–2013)," at www.un.org/en/peacekeeping/resources/statistics/contributors_archive.shtml.

9. Ibid.

10. UNDPI, op. cit. note 1.

11. Ibid.

12. UNDPI, "United Nations Political and Peacebuilding Missions. Fact Sheet: 31 January 2014" (New York: February 2014).

13. Ibid.

14. UNDPKO, "Fatalities," 6 March 2014, at www.un.org/en/peacekeeping/resources/statistics/fatalities.shtml.

15. Ibid.

16. UNDPI, op. cit. note 1.

17. Ibid.

18. Ibid.

19. Ibid.

20. Ibid.

21. Ibid.

22. Ibid.

23. Calculated from UNDPKO, "Monthly Summary of Contributions (Police, UN Military Experts on Mission and Troops). As of 31 January 2014," at www.un.org/en/peacekeeping/contributors/2014/jan14_1.pdf.

24. Calculated from ibid.

25. Ibid.

26. Ibid.

27. UNDPKO, "Financing Peacekeeping," at www.un.org/en/peacekeeping/operations/financing.shtml.

28. Ibid.

29. Year-end figures. The United Nations reports arrears in current dollar terms, but data here are in 2013 dollars.

30. UNDPI, op. cit. note 1.

31. Zentrum für Internationale Friedenseinsätze (ZIF), "International and German Personnel in Peace Operations 2013–14," Berlin, September 2013.

32. International Security Information Service, "CSDP Note: Chart and Table of CSDP and EU missions," Brussels, March 2014.

33. ZIF, op. cit. note 31.

34. Ibid.

35. Ibid.

36. Ibid.

37. UNDPI, "United Nations Peacekeeping Operations. Fact Sheet: 31 December 2013" (New York: January 2014); UNDPI, op. cit. note 12; ZIF, op. cit. note 31.

38. UNDPI, op. cit. note 37; UNDPI, op. cit. note 12; ZIF, op. cit. note 31.

39. UNDPI, op. cit. note 37; UNDPI, op. cit. note 12; ZIF, op. cit. note 31.

Displaced Populations (pages 96–99)

1. UN High Commissioner for Refugees (UNHCR), *Global Trends 2012* (Geneva: 2013); UN Relief and Works Agency for Palestine Refugees in the Near East (UNRWA), "Statistics," at www.unrwa.org/etemplate.php?id=253; Internal Displacement Monitoring Centre (IDMC), *Global Overview 2012: People Internally Displaced by Conflict and Violence* (Geneva: 2013); IDMC, *Global Estimates 2012: People Displaced by Disasters* (Geneva: 2013).

2. UNHCR, op. cit. note 1.

3. UNRWA, op. cit. note 1.

4. IDMC, *People Internally Displaced by Conflict and Violence*, op. cit. note 1, p. 10.

5. IDMC, *People Displaced by Disasters*, op. cit. note 1, p. 11.

6. International Federation of Red Cross and Red Crescent Societies (IFRC), *World Disasters Report 2012* (Geneva: 2012), p. 15.

7. Total displaced from UNHCR, op. cit. note 1, from UNRWA, op. cit. note 1, from IDMC, *People Internally Displaced by Conflict and Violence*, op. cit. note 1, and from IDMC, *People Displaced by Disasters*, op. cit. note 1.

8. IFRC, op. cit. note 6.

9. Ibid., p. 14.

10. UNHCR, *The State of the World's Refugees 2000: Fifty Years of Humanitarian Action* (New York: Oxford University Press, 2000), Annex 3; UNHCR, *Global Trends*, 2008–11 editions (Geneva: 2009–12); UNRWA, op. cit. note 1; IDMC, *People Internally Displaced by Conflict and Violence*, op. cit. note 1, p. 10; IDMC, "Global IDP Estimates (1990–2011)," at www.internal-displacement.org/8025708F004CE90B/(http Pages)/10C43F54DA2C34A7C12573A1004EF9FF?Open Document&count=1000.

11. UNHCR, *The State of the World's Refugees 2000*, op. cit. note 10; UNHCR, *Global Trends*, op. cit. note 10.

12. UNRWA, op. cit. note 1.

13. IDMC, op. cit. note 10.

14. UNHCR, op. cit. note 1, p. 3.

15. Ibid., p. 14.

16. Ibid., p. 15.

17. Ibid., pp. 2–3, 11.

18. Ibid., pp. 2, 5.

19. Ibid., pp. 3, 11.

20. IDMC, *People Internally Displaced by Conflict and Violence*, op. cit. note 1, p. 8.

21. Ibid.

22. Ibid.

23. UNHCR, *Global Trends 2011* (Geneva: 2012), p. 7; UNHCR, op. cit. note 1, p. 5.

24. UNHCR, *The State of the World's Refugees 2012: In Search of Solidarity* (Geneva: 2012), p. 5.

25. IDMC, *People Displaced by Disasters*, op. cit. note 1, p. 11.

26. Ibid., pp. 25–26.

27. Ibid., p. 25.

28. Ibid., p. 31.

29. Ibid., p. 33.

30. Ibid., p. 34.

31. Ibid., p. 13.

32. Ibid., p. 12.

33. Ibid.

34. International Organization for Migration, cited in Actionaid et al., *Into Unknown Territory: The Limits to Adaptation and Reality of Loss and Damage from Climate Impacts* (Bonn: 2012), p. 7.

35. Ibid., p. 9.

36. UNHCR, op. cit. note 24, p. 2.

37. Ibid.

38. IFRC, op. cit. note 6.

39. Christian Aid, *Human Tide: The Real Migration Crisis* (London: 2007), p. 5.

40. IFRC, op. cit. note 6, p. 148.

41. Ibid., p. 18.

World Population: Fertility Surprise Implies More Populous Future (pages 100–04)

1. U.N. Population Division, *World Population Prospects:*

The 2012 Revision, Population Database, at esa.un.org/wpp /unpp/panel_population.htm, viewed 19 June 2013.

2. Ibid.

3. Ibid.

4. Ibid.

5. Ibid.

6. Ibid.

7. U.N. Population Division, *World Population Prospects: The 2002 Revision*, at www.un.org/esa/population/publica tions/wpp2002/WPP2002-HIGHLIGHTSrev1.PDF.

8. U.N. Population Division, op. cit. note 1.

9. Ibid.

10. "UN: World Population to Reach 8.1 Billion in 2025, UN Says," *CBC News*, 13 June 2013.

11. Earlier fertility projections from U.N. Population Division, *World Population Prospects: The 2010 Revision, Volume I: Comprehensive Tables*, ST/ESA/SER.A/313, at www.esa.un.org /unpd/wpp/Documentation/pdf/WPP2010_Volume-I _Comprehensive-Tables.pdf, viewed 2 July 2013; later projections from U.N. Population Division, op. cit. note 1.

12. U.N. Population Division, op. cit. note 1.

13. U.N. Population Division, op. cit. note 11; U.N. Population Division, op. cit. note 1.

14. U.N. Population Division, "World Population Projected to Reach 9.6 Billion by 2050 with Most Growth in the Developing Regions, Especially Africa—Says UN," press release (New York: 13 June 2013).

15. Margaret Greene, Shareen Joshi, and Omar Robles, *The State of World Population 2012: By Choice, Not By Chance— Family Planning, Human Rights and Development* (New York: U.N. Population Fund, 2012), p. 90.

16. Susheela Singh and Jacqueline E. Darroch, *Adding It Up: Costs and Benefits of Contraceptive Services, Estimates for 2012* (New York: Guttmacher Institute, 2012).

17. Ibid.

18. Alex C. Ezeh, Blessing U. Mberu, and Jacques O. Emina, "Stall in Fertility Decline in Eastern African Countries: Regional Analysis of Patterns, Determinants and Implications," *Philosophical Transactions of the Royal Society B*, October 2009, pp. 2991–3007; Susheela Singh et al., *Abortion Worldwide: A Decade of Uneven Progress* (New York: Guttmacher Institute, 2009), Appendix Tables 3 and 4.

19. Singh et al., op. cit. note 18.

20. U.N. Population Division, op. cit. note 1.

21. Ibid.

22. Ibid.

23. Ibid.

24. Ibid.

25. Ibid.

26. Ibid.

27. Ibid.

28. Arthur Haupt, Thomas T. Kane, and Carl Haub, *PRB's Population Handbook, 6th ed.* (Washington, DC: Population Reference Bureau, 2011).

29. Anthony Faiola, "Portugal's Birthrate Plummeting, a Sign of More Economic Troubles Ahead," *Washington Post*, 23 June 2013.

30. U.N. Population Division, op. cit. note 1.

31. Ibid.

32. Ibid.

33. Ibid.

34. Ibid.

35. Ibid.

36. Ibid.

37. Ibid.

38. Ibid.

39. Ibid.

40. Ibid.

41. Ibid.

42. Ibid.

43. Ibid.

44. Ibid.

45. Ibid.

46. Ibid.

47. Ibid.

48. Ibid.

49. Ibid.

50. Ibid.

51. Ibid.

52. Ibid.

53. Ibid.

54. Ibid.

55. Ibid.

56. Haupt, Kane, and Haub, op. cit note 28.

57. Data from U.N. Population Division, op. cit. note 1.

58. Ibid.

59. Ibid.

60. Ibid.

61. Ibid.

62. Ibid.

63. Ibid.; Martin L. Parry et al., *Climate Change 2007: Impacts, Adaptation and Vulnerability—Contribution of Working Group II to the Fourth Assessment Report of the Intergovernmental Panel on Climate Change* (Cambridge, U.K.: Cambridge University Press, 2007); Peter H. Gleick et al., *The World's Water, Vol. 7* (Washington, DC: Island Press, 2012).

64. U.N. Population Division, op. cit. note 1.

65. Ibid.

66. Parry et al., op. cit. note 63; Gleick et al., op. cit. note 63.

Women as National Legislators (pages 105–07)

1. Inter-Parliamentary Union (IPU), "Women in National Parliaments," at www.ipu.org/wmn-e/world.htm, viewed 27 January 2014.

2. Ibid.

3. Ibid.

4. IPU, "World Classification" (table), in IPU, op. cit. note 1.

5. Ibid.

6. Ibid.

7. "Rwanda: Women Win 64 Percent of Seats in Parliamentary Elections, Maintaining Number One Spot Worldwide," *All Africa*, 23 September 2013 Online edition.

8. Drude Dahlerup, "Global Database of Quotas for Women," based on Drude Dahlerup, "Increasing Women's Political Representation: New Trends in Gender Quotas," in Julie Ballington and Azza Karam, eds., *Women in Parliament: Beyond Numbers. A Revised Edition* (Stockholm: International IDEA, 2005); Drude Dahlerup, ed., *Women, Quotas and Politics* (New York: Routledge, 2006).

9. IPU, op. cit. note 4.

10. UN Women calculation based on IDEA, Stockholm University, and IPU, "Global Data Base of Quotas on Women," at www.unwomen.org/en/what-we-do/leadership-and-political-participation/facts-and-figures#sthash.63LSmH0g.dpuf, viewed June 2013, and on IPU, op. cit. note 1.

11. Ibid.

12. Poornima and Vinod Vyasulu, "Women in Panchayati Raj: Grassroots Democracy in India, Experience from Malgudi," Background Paper for U.N. Development Programme, Meeting on Women and Political Participation: 21st Century Challenges, New Delhi, 24–26 March 1999.

13. Clara Araújo, "The Implementation of Quotas: Latin American Experiences," Department of Social Sciences, State University of Rio de Janeiro, Brazil, 24 February 2003.

14. Dahlerup, "Increasing Women's Political Representation," op. cit. note 8.

15. R. Chattopadhyay and E. Duflo, "Women as Policy Makers: Evidence from a Randomized Policy Experiment in India," *Econometrica*, vol. 72, no. 5 (2004), pp. 1409–43.

16. K. A. Bratton and L. P. Ray, "Descriptive Representation: Policy Outcomes and Municipal Day-Care Coverage in Norway," *American Journal of Political Science*, vol. 46, no. 2 (2002), pp. 428–37.

17. Baha'i International Community, "Traditional Media as Change Agent," *One Country* (newsletter), October–December 1993, cited in Augusto Lopez-Claros and Saadia Zahidi, *Women's Empowerment: Measuring the Global Gender Gap* (Geneva: World Economic Forum, 2006).

18. "Progress of the World's Women 2000. Biennial Report," cited in World Economic Forum, op. cit. note 17.

19. Jodi L. Jacobson, *Gender Bias: Roadblock to Sustainable Development*, Worldwatch Paper 110 (Washington, DC: Worldwatch Institute, September 1992), cited in Molly O. Sheehan, "Women Slowly Gain Ground in Politics," in Worldwatch Institute, *Vital Signs 2000* (New York: W. W. Norton & Company, 2000), p. 152.

20. Sheehan, op. cit. note 19; IPU, op. cit. note 1.

21. Ibid.

22. "Women's Day in Kuwait," *The Economist*, 22 May 1999; Douglas Jehl, "Debate on Women's Rights Shows Deep Rift in Kuwait Society," *New York Times*, 20 December 1999, cited in Sheehan, op. cit. note 19.

23. UNICEF calculations based on data from the IPU, op. cit. note 1, cited in UNICEF, *State of the World's Children 2007* (New York: 2006), pp. 51–62.

24. The Convention on the Elimination of All Forms of Discrimination Against Women (CEDAW), at www.cedaw2012.org/index.php/about-cedaw.

25. Ibid.

26. Women 2000, "Beijing +5 at a Glance," New York, 5–9 June 2000, at www.un.org/womenwatch/confer/beijing5.

27. UN Women, "Platform for Action," Fourth World Con-

ference on Women, at www.un.org/womenwatch/daw/beijing/platform.

Mobile Phone Growth Slows as Mobile Devices Saturate the Market (pages 108–11)

1. A.T. Kearney and GSMA, *The Mobile Economy 2013* (London: 2013).

2. U.N. Department of Economic and Social Affairs, *World Population Prospects: The 2012 Revision* (New York: 2012).

3. InfoDev and World Bank, *Infographic: Maximizing Mobile for Development* (Washington, DC: 2012).

4. Figure for 2000 from InfoDev and World Bank, *Information and Communications for Development 2012: Maximizing Mobile* (Washington, DC: 2012); 2013 data from International Telecommunication Union (ITU), *Measuring the Information Society 2013* (Geneva: 2013).

5. A.T. Kearney and GSMA, op. cit. note 1.

6. ITU, "ITU Releases Latest Global Technology Development Figures," press release (Geneva: 27 February 2013).

7. GSMA, *Scaling Mobile for Development: A Developing World Opportunity* (2013).

8. ITU, *Statistics* (Geneva: 2012).

9. Ibid.

10. Ibid.

11. Ibid.

12. Ibid.

13. GSMA, op. cit. note 7.

14. ITU, op. cit. note 8.

15. ITU, *Measuring the Information Society 2012* (Geneva: 2012).

16. Malcolm Foster, "Cell Phones Vital in Developing World," *Associated Press*, 27 January 2007.

17. ITU, op. cit. note 4.

18. ITU, *The World in 2013: ICT Facts and Figures* (Geneva: 2013).

19. Ibid.

20. Ericsson, *Ericsson Mobility Report on the Pulse of the Networked Society* (2012).

21. Darrell West, "Alleviating Poverty: Mobile Communications, Microfinance and Small Business Development Around the World," Center for Technology Innovation, Brookings Institution, Washington, DC, 16 May 2013.

22. Ibid.

23. Mark Cohen, "Text-Message Marketing," *New York Times*, 23 September 2009.

24. Courtney E. Martin, "Medicine by Text Message: Learning from the Developing World," *The Atlantic*, 4 April 2013.

25. Ibid.

26. Olivia Solon, "MFarm Empowers Kenya's Farmers with Price Transparency and Market Access," *Wired Magazine*, June 2013.

27. InfoDev and World Bank, op. cit. note 4.

28. Ibid.

29. Carol Huang, "Facebook and Twitter Key to Arab Spring Uprisings: Report," *The National*, 6 June 2011.

30. InfoDev and World Bank, op. cit. note 4.

31. Ibid.

32. Office of the High Commissioner for Human Rights, *Democratic Republic of the Congo 1993-2003, UN Mapping Report: Violence Linked to Natural Resource Exploitation* (Geneva: 2010).

33. Ibid.

34. Ibid.

35. U.N. Security Council, "Letter Dated 29 November 2011 from the Chair of the Security Council Committee Established Pursuant to Resolution 1533 (2004) Concerning the Democratic Republic of the Congo Addressed to the President of the Security Council," New York, 2011.

36. China Labor Watch, *Beyond Foxconn: Deplorable Working Conditions Characterize Apple's Entire Supply Chain* (New York: 2012).

37. Charles Duhigg and David Barboza, "In China, Human Costs are Built Into an iPad," *New York Times*, 25 January 2012.

38. China Labor Watch, *An Investigation of Eight Samsung Factories in China* (New York: 2012).

39. Leyla Acaroglu, "Where do Old Cellphones Go to Die?" *New York Times*, 4 May 2013.

40. Ibid.

41. Ibid.

The Vital Signs Series

Some topics are included each year in *Vital Signs*; others are covered only in certain years. The following is a list of topics covered in *Vital Signs* thus far, with the year or years they appeared indicated in parentheses. The reference to 2006 indicates *Vital Signs 2006–2007*; 2007 refers to *Vital Signs 2007–2008*. The year 2013 indicates *Volume 20*, and 2014 is this edition, *Volume 21*.

ENERGY AND TRANSPORTATION
Fossil Fuels
Carbon Use (1993)

Coal (1993–96, 1998, 2009, 2011)

Coal and Natural Gas Combined (2013)

Fossil Fuels Combined (1997, 1999–2003, 2005–07, 2010, 2014)

Natural Gas (1992, 1994–96, 1998, 2011–12)

Oil (1992–96, 1998, 2009, 2012–13)

Renewables, Efficiency, Other Sources
Biofuels (2005–07, 2009–12, 2014)

Biomass Energy (1999)

Combined Heat and Power (2009)

Compact Fluorescent Lamps (1993–96, 1998–2000, 2002, 2009)

Efficiency (1992, 2002, 2006)

Geothermal Power (1993, 1997)

Hydroelectric Power (1993, 1998, 2006, 2012)

Hydropower and Geothermal Combined (2013)

Nuclear Power (1992–2003, 2005–07, 2009, 2011–12, 2014)

Smart Grid (2013)

Solar Power (1992–2002, 2005–07, 2009–12)

Solar and Wind Power (2014)

Solar Thermal Power (2010)

Wind Power (1992–2003, 2005–07, 2009–13)

Transportation
Air Travel (1993, 1999, 2005–07, 2011, 2014)

Bicycles (1992–2003, 2005–07, 2009)

Car-sharing (2002, 2006)

Electric Cars (1997)

Gas Prices (2001)

High-Speed Raid (2012)

Motorbikes (1998)

Railroads (2002)

Urban Transportation (1999, 2001)

Vehicles (1992–2003, 2005–07, 2009–14)

ENVIRONMENT AND CLIMATE
Atmosphere and Climate
Agriculture as Source of Greenhouse Gases (2014)

Carbon and Temperature Combined (2003, 2005–07, 2009–10)

Carbon Capture and Storage (2012–13)

Carbon Emissions (1992, 1994–2002, 2009, 2013–14)

CFC Production (1992–96, 1998, 2002)

Global Temperature (1992–2002)

Ozone Layer (1997, 2007)

Sea Level Rise (2003, 2011)

Weather-related Disasters (1996–2001, 2003, 2005–07, 2009–11, 2013–14)

Natural Resources, Animals, Plants
Amphibians (1995, 2000)

Aquatic Species (1996, 2002)

Birds (1992, 1994, 2001, 2003, 2006)

Coral Reefs (1994, 2001, 2006, 2010)

Dams (1995)

Ecosystem Conversion (1997)

Energy Productivity (1994, 2012)

Forests (1992, 1994–98, 2002, 2005–06, 2012)

Groundwater (2000, 2006)

Ice Melting (2000, 2005)

Invasive Species (2007)

Mammals (2005)

Mangroves (2006)

Marine Mammals (1993)

Organic Waste Reuse (1998)
Plant Diversity (2006)
Primates (1997)
Terrestrial Biodiversity (2007, 2011)
Threatened Species (2007)
Tree Plantations (1998)
Vertebrates (1998)
Water Scarcity (1993, 2001–02, 2010, 2013)
Water Tables (1995, 2000)
Wetlands (2001, 2005)

Pollution

Acid Rain (1998)
Air Pollution (1993, 1999, 2005)
Algal Blooms (1999)
Hazardous Wastes (2002)
Lead in Gasoline (1995)
Mercury (2006)
Nuclear Waste (1992, 1995)
Ocean (2007)
Oil Spills (2002)
Pollution Control Markets (1998)
Sulfur and Nitrogen Emissions (1994–97)

Other Environmental Topics

Bottled Water (2007, 2011)
Energy Poverty (2012)
Environmental Indicators (2006)
Environmental Treaties (1995, 1996, 2000, 2002)
Protected Areas (2010)
Semiconductor Impacts (2002)
Transboundary Parks (2002)
World Heritage Sites (2003)

FOOD AND AGRICULTURE

Agriculture

Agricultural Population (2014)
Farm Animals (2013)
Farmland Quality (2002)
Fertilizer Use (1992–2001, 2011)
Foreign Investment in Farmland (2013)
Genetically Modified Crops (1999–2002, 2009)
Grain Area (1992–93, 1996–97, 1999–2000)
Irrigation (1992, 1994, 1996–99, 2002, 2007, 2010, 2013)
Nitrogen Fixation (1998)
Organic Agriculture (1996, 2000, 2010, 2012–13)

Pesticide Control or Trade (1996, 2000, 2002, 2006)
Pesticide Resistance (1994, 1999)
Soil Erosion (1992, 1995)
Urban Agriculture (1997)
Women Farmers (2013)

Food Trends

Aquaculture (1994, 1996, 1998, 2002, 2005, 2013)
Aquaculture and Fish Harvest Combined (2006–07, 2009–12)
Cocoa Production (2002, 2011)
Coffee (2001)
Eggs (2007)
Fish Harvest (1992–2000)
Grain Production (1992–2003, 2005–07, 2009–13)
Grain Stocks (1992–99)
Grain Used for Feed (1993, 1995–96)
Livestock (2001)
Meat (1992–2000, 2003, 2005–07, 2009–13)
Milk (2001)
Soybeans (1992–2001, 2007)
Sugar and Sweetener Use (2002, 2012)

GLOBAL ECONOMY AND RESOURCES

Resource Economics

Agricultural Subsidies (2003, 2014)
Aluminum (2001, 2006–07)
Arms and Grain Trade (1992)
Commodity Prices (2001, 2014)
Fossil Fuel Subsidies (1998, 2012–14)
Gold (1994, 2000, 2007)
Illegal Drugs (2003)
Materials Use (2011)
Metals Exploration (1998, 2002)
Metals Production (2002, 2010, 2013)
Municipal Solid Waste (2013)
Paper (1993–94, 1998–2000)
Paper Recycling (1994, 1998, 2000)
Payments for Ecosystem Services (2012)
Renewable Energy Investments (2013)
Renewable Energy Policy (2014)
Roundwood (1994, 1997, 1999, 2002, 2006–07, 2011)
Steel (1993, 1996, 2005–07)

Landmines (1996, 2002)

Military Expenditures (1992, 1998, 2003, 2005–06, 2014)

Nuclear Arsenal (1992–96, 1999, 2001, 2005, 2007)

Peacekeeping Expenditures (1994–2003, 2005–07, 2009, 2014)

Resource Wars (2003)

Wars (1995, 1998–2003, 2005–07)

Small Arms (1998–99)

Reproductive Health and Women's Status

Family Planning Access (1992)

Female Education (1998)

Fertility Rates (1993)

Gender Gap (2012)

Maternal Mortality (1992, 1997, 2003)

Population Growth (1992–2003, 2005–07, 2009–11, 2014)

Sperm Count (1999, 2007)

Violence Against Women (1996, 2002)

Women in Politics (1995, 2000, 2014)

Other Social Topics

Aging Populations (1997)

Climate Change Migration (2013)

Co-operatives (2013)

Educational Levels (2011)

Homelessness (1995)

Income Distribution or Poverty (1992, 1995, 1997, 2002–03, 2010)

Language Extinction (1997, 2001, 2006)

Literacy (1993, 2001, 2007)

International Criminal Court (2003)

Millennium Development Goals (2005, 2007)

Nongovernmental Organizations (1999)

Orphans Due to AIDS Deaths (2003)

Prison Populations (2000)

Public Policy Networks (2005)

Quality of Life (2006)

Refugees (1993–2000, 2001, 2003, 2005, 2014)

Refugees-Environmental (2009)

Religious Environmentalism (2001)

Slums (2006)

Social Security (2001)

Sustainable Communities (2007)

Teacher Supply (2002)

Urbanization (1995–96, 1998, 2000, 2002, 2007, 2013)

Voter Turnouts (1996, 2002)